Springer Theses

Recognizing Outstanding Ph.D. Research

Aims and Scope

The series "Springer Theses" brings together a selection of the very best Ph.D. theses from around the world and across the physical sciences. Nominated and endorsed by two recognized specialists, each published volume has been selected for its scientific excellence and the high impact of its contents for the pertinent field of research. For greater accessibility to non-specialists, the published versions include an extended introduction, as well as a foreword by the student's supervisor explaining the special relevance of the work for the field. As a whole, the series will provide a valuable resource both for newcomers to the research fields described, and for other scientists seeking detailed background information on special questions. Finally, it provides an accredited documentation of the valuable contributions made by today's younger generation of scientists.

Theses are accepted into the series by invited nomination only and must fulfill all of the following criteria

- They must be written in good English.
- The topic should fall within the confines of Chemistry, Physics, Earth Sciences, Engineering and related interdisciplinary fields such as Materials, Nanoscience, Chemical Engineering, Complex Systems and Biophysics.
- The work reported in the thesis must represent a significant scientific advance.
- If the thesis includes previously published material, permission to reproduce this must be gained from the respective copyright holder.
- They must have been examined and passed during the 12 months prior to nomination.
- Each thesis should include a foreword by the supervisor outlining the significance of its content.
- The theses should have a clearly defined structure including an introduction accessible to scientists not expert in that particular field.

More information about this series at http://www.springer.com/series/8790

Mona Alimohammadi

Aortic Dissection: Simulation Tools for Disease Management and Understanding

Doctoral Thesis accepted by
University College London, London, UK

Author
Dr. Mona Alimohammadi
Department of Mechanical Engineering
University College London
London
UK

Supervisors
Dr. Vanessa Diaz
Department of Mechanical Engineering
University College London
London
UK

Dr. Stavroula Balabani
Department of Mechanical Engineering
University College London
London
UK

I, Mona Alimohammadi, confirm that the work presented in this thesis is my own. Where information has been derived from other sources, I confirm that this has been indicated in the thesis.

ISSN 2190-5053   ISSN 2190-5061 (electronic)
Springer Theses
ISBN 978-3-319-85886-9   ISBN 978-3-319-56327-5 (eBook)
https://doi.org/10.1007/978-3-319-56327-5

© Springer International Publishing AG 2018
Softcover re-print of the Hardcover 1st edition 2018
This work is subject to copyright. All rights are reserved by the Publisher, whether the whole or part of the material is concerned, specifically the rights of translation, reprinting, reuse of illustrations, recitation, broadcasting, reproduction on microfilms or in any other physical way, and transmission or information storage and retrieval, electronic adaptation, computer software, or by similar or dissimilar methodology now known or hereafter developed.
The use of general descriptive names, registered names, trademarks, service marks, etc. in this publication does not imply, even in the absence of a specific statement, that such names are exempt from the relevant protective laws and regulations and therefore free for general use.
The publisher, the authors and the editors are safe to assume that the advice and information in this book are believed to be true and accurate at the date of publication. Neither the publisher nor the authors or the editors give a warranty, express or implied, with respect to the material contained herein or for any errors or omissions that may have been made. The publisher remains neutral with regard to jurisdictional claims in published maps and institutional affiliations.

Printed on acid-free paper

This Springer imprint is published by Springer Nature
The registered company is Springer International Publishing AG
The registered company address is: Gewerbestrasse 11, 6330 Cham, Switzerland

To my parents

Supervisors' Foreword

The Ph.D. thesis of Mona Alimohammadi describes a new paradigm in cardiovascular interventional planning using aortic dissections as an exemplar.

Aortic Dissection (AD), is a rare, life threatening cardiovascular disease. The thesis focuses on type-B aortic dissection, which affects the descending aorta and is associated with a large number of complications, such as vessel wall rupture, thrombosis or malperfusion. These are patient-specific and preclude a 'one-fits-all' treatment approach. Clinical decisions regarding diagnosis, management and treatment of AD are incredibly difficult; aortic dissection is not diagnosed on its initial presentation in 15–43% of cases. Initial management of diagnosed or highly suspected acute aortic dissection focuses on pain control, heart rate and then blood pressure management, and immediate surgical consultation (typically by considering whether or not stenting the upper tear). It is an ever-increasing problem; when the condition is identified, therapy is driven by a set of key principles and plagued by a number of management pitfalls.

To aid clinicians to tailor treatment to individual patient needs, additional information on the haemodynamic environment in the diseased aorta is required, which is not readily available from routine clinical tests. However, combining routine clinical data such as CT scans and pressure measurements with computational modelling can provide a powerful tool to evaluate the efficacy of various treatment options and aid clinical decision making, as this thesis demonstrates.

The thesis makes three key contributions to this modelling paradigm: (1) appropriate definition of patient-specific boundary conditions, based on routine clinical data and Windkessel models, (2) demonstration of a virtual surgery approach to treat aortic dissection and (3) the first simulations of aortic dissection considering the influence of vessel wall motion and intimal flap dynamics on the haemodynamics of the disease.

The study provides a strong foundation for further developments in the field and represents a significant step towards personalised medicine. Mona was awarded the IMechE Best Medical Engineering Thesis Prize for this work, and we are very pleased to see the thesis being published in the Springer theses series.

London, UK
March 2017

Dr. Vanessa Diaz
Dr. Stavroula Balabani

Abstract

Aortic dissection is a severe cardiovascular pathology in which a tear in the intimal layer of the aortic wall allows blood to flow between the vessel wall layers, forming a 'false lumen'. In type-B aortic dissections, those involving only the descending aorta, the decision to medically manage or surgically intervene is not clear and is highly dependent on the patient. In addition to clinical imaging data, clinicians would benefit greatly from additional physiological data to inform their decision-making process. Computational fluid dynamics methods show promise for providing data on haemodynamic parameters in cardiovascular diseases, which cannot otherwise be predicted or safely measured. The assumptions made in the development of such models have a considerable impact on the accuracy of the results, and thus require careful investigation. Application of appropriate boundary conditions is a challenging but critical component of such models. In the present study, imaging data and invasive pressure measurements from a patient with a type-B aortic dissection were used to assist numerical modelling of the haemodynamics in a dissected aorta. A technique for tuning parameters for coupled Windkessel models was developed and evaluated. Two virtual treatments were modelled and analysed using the developed dynamic boundary conditions. Finally, the influence of wall motion was considered, of which the intimal flap that separates the false lumen from the true lumen, is of particular interest. The present results indicate that dynamic boundary conditions are necessary in order to achieve physiologically meaningful flows and pressures at the boundaries, and hence within the dissected aorta. Additionally, wall motion is of particular importance in the closed regions of the false lumen, wherein rigid wall simulations fail to capture the motion of the fluid due to the elasticity of the vessel wall and intimal flap.

List of Publications

Journal Publications

1. Alimohammadi, M., Agu, O., Balabani, S. and Díaz-Zuccarini, V. (2014), 'Development of a patient-specific tool to analyse aortic dissections: Assessment of mixed patient-specific flow and pressure boundary condition', *Medical Engineering and Physics* 36, 275–284.
2. Alimohammadi, M., Bhattacharya-Ghosh, B., Seshadri, S., Penrose, J., Agu, O., Balabani, S. and Díaz-Zuccarini, V. (2014), 'Evaluation of the hemodynamic effectiveness of aortic dissection treatments via virtual stenting', *International Journal of Artificial Organs* 37(10), 753–762.
3. Decorato, I., Salsac, A.-V., Legallais, C., Alimohammadi, M., Díaz-Zuccarini, V. and Kharboutly, Z. (2014), 'Influence of an arterial stenosis on the hemodynamics within an arteriovenous fistula (AVF): comparison before and after balloon-angioplasty', *Cardiovascular Engineering and Technology* 5(3), 233–243.
4. Alimohammadi, M., Pichardo-Almarza, C., Di Tomaso, G., Balabani, S., Agu, O. and Díaz-Zuccarini, V. (2015), 'Predicting atherosclerotic plaque location in an iliac bifurcation using a hybrid CFD/biomechanical approach', *in IWBBIO 2015, LNCS 9044*, 594–606.
5. Alimohammadi, M., Sherwood, J.M., Karimpour, M., Agu, O., Balabani, S. and Díaz-Zuccarini, V 'Aortic dissection simulation models for clinical support: fluid-structure interaction vs. rigid wall models'. *Biomedical Engineering OnLine*. 14:34.

Book Chapters

Díaz-Zuccarini, V., Alimohammadi, M., Pichardo-Almarza, C., (2015), Reflecting on DISCIPULUS and Remaining Challenges, in 'The Digital Patient: Advancing Medical Research, Education, and Practice', Eds. Combs, C.D., Sokolowski, J.A., Banks, C.M., *John Wiley and Sons*.

Conferences

Presentations: M Alimohammadi et al., 'Coupled 0D and CFD simulation of blood flow through a Stanford type-B aortic dissection', *Congress of the European Society of Biomechanics*, Lisbon, 1–4 July, 2012.
M Alimohammadi et al., 'Coupled 0D and CFD simulation of blood flow through a Stanford type-B aortic dissection', *IMA Mathematics of Medical Devices and Surgical Procedures*, London, 17–19 September, 2012.

Posters: M Alimohammadi et al., 'Fluid structure interaction simulation of blood flow through a patient-specific Stanford type-B aortic dissection', *World Congress of Biomechanics*, Boston, July 6–11, 2014.

M Alimohammadi et al., 'Coupled 0D and CFD simulation of blood flow through a Stanford type-B aortic dissection', *IMA Mathematics of Medical Devices and Surgical Procedures*, London, 17–19 September, 2012. Winner of first prize graduate student poster competition.

Workshops Attended

euHeart Project workshop; euHeart is the integrated cardiac care using patient-specific cardiovascular modelling. The workshop focused on the numerical and computational models of aortic coarctation and aortic aneurysm. Sheffield University, 17–18 October, 2011.

London Mathematical Society/EPSRC Course on Continuum Mechanics in Biology and Medicine. University College London, 17–22 June, 2012

Acknowledgements

I was lucky enough to have two excellent supervisors to whom I am indebted.

I would like to sincerely thank Dr. Vanessa Diaz for accepting me as her Ph.D. student and it has been a privilege to work with her during the past four years. She has always been generous with her time, and greatly motivated me to work in this fascinating field of physiological modelling. She supported me selflessly and gave me confidence when I needed it during critical times. Her guidance and forward thinking always kept me on track. I will always be grateful to her for introducing me to dynamic boundary conditions and the wonders of FORTRAN.

I would like to give my gratitude to Dr. Stavroula Balabani for her co-supervision. She has been exceedingly patient, kind and supportive of my work, and enthusiastic about exploring new ideas. Her lectures and our many discussions during this project have developed my love for fluid dynamics. I am continually amazed by her ability to recall specifics of papers and the fastidiousness of her reviewing approach. I am privileged to be her first female Ph.D. student.

I would like to thank Mr. Obi Agu, my clinical supervisor for his time and effort in teaching me about aortic dissection and providing invaluable clinical data. I very much appreciate him taking me to the operation theatre to give me insight into the true complexity involved in the treatment of patients suffering from aortic dissection.

I have great respect and appreciation to Dr. Cesar Pichardo for humbly sharing and motivating me.

I am deeply grateful to my good friend, Dr. Morad Karimpour for his help and support and for being there for me, despite the time difference.

My thanks go to my kind friend Dr. John Vardakis for the useful discussions and encouragement.

I would like to thank all of the staff in the department of mechanical engineering for their support and assistance, as well as Tristan Clark and Denis Timm for IT management.

Dr. George Papadakis must also receive my thanks for introducing me to numerical methods and CFD. I very much enjoyed his lectures and am grateful for his supervision during my MRes project, which led me to this work.

I must also give my thanks and respect to all of the individuals who have, and will, participate in studies such as this, without which this field of research would not be possible.

I feel honoured to have amazing friends, who were always eager to get updates about my progress and supported me to achieve more.

I would like to sincerely thank Dr. Joseph Sherwood for sharing with me his indefatigable passion for haemodynamics. I enjoyed all the useful discussions we had, which inspired me to explore and learn more. I also greatly appreciate his care and support during difficult times and I am so happy that he is now using lumped parameter models and CFD in his research.

Finally, I would like to thank my family for their love, support and patience. Baba joonam, this journey started because of you and only now can I see what you saw, so thanks for your wisdom and foresight. Maman joonam, thanks for the time and effort you gave for my education, I hope this has put a smile on your face and I am glad you no longer need to worry about me running away from schools. My dear Tina, no, you don't have to do a Ph.D. (but I bet you will), and thanks for all those intellectual talks and support when I needed it the most. Mina, you don't have to do a Ph.D. either, as you have already shared the emotional side with me. Thank you for supporting, tolerating and encouraging me.

Contents

1	**Introduction**		1
	1.1 Motivation and Background		1
		1.1.1 Introduction to the Circulatory System	2
		1.1.2 Aortic Dissection	5
	1.2 Numerical Modelling of the Cardiovascular System		16
		1.2.1 Background	16
		1.2.2 Approaches to Modelling Aortic Dissection	17
		1.2.3 Objectives of Aortic Dissection Modelling	25
	1.3 Objectives of the Present Research		27
	1.4 Outline of the Thesis		28
	References		29
2	**Computational Methods for Patient-Specific Modelling**		39
	2.1 Computational Fluid Dynamics		39
		2.1.1 Governing Equations	40
		2.1.2 Numerical Implementation	41
	2.2 Building the Fluid Domain		43
		2.2.1 Introduction to Clinical Imaging	43
		2.2.2 3D Domain Extraction	44
		2.2.3 Geometry	48
	2.3 Meshing		51
	2.4 Dynamic Boundary Conditions		52
		2.4.1 Analogue Equations	52
		2.4.2 Two-element Windkessel Model	53
		2.4.3 Three-element Windkessel Model	54
		2.4.4 Four-element Windkessel Models	56
		2.4.5 Compound Windkessel Models	57
		2.4.6 Parameters for Windkessel Models	57
		2.4.7 Comparison of Zero-Pressure and Windkessel Boundary Conditions for Aortic Dissection	59

		2.4.8	Comparison of Flow-Split and Windkessel Boundary Conditions for Aortic Dissection	61
	2.5	Finite Element Modelling		62
		2.5.1	Vessel Wall Reconstruction	63
		2.5.2	Material Properties	63
	References			65

3 Haemodynamics of a Dissected Aorta 69
3.1 Introduction 69
3.2 Methodology 70
3.2.1 Boundary Conditions and Data Assimilation Method 71
3.3 Results 75
3.3.1 Flow Characteristics 75
3.3.2 Pressure Distribution 79
3.3.3 Wall Shear Stress 81
3.4 Discussion 84
3.4.1 Limitations 85
3.5 Sensitivity of the Windkessel Parameters 87
3.6 Mesh Sensitivity 89
3.7 Effect of Turbulence Modelling 95
3.8 Conclusions 97
References 97

4 Effectiveness of Aortic Dissection Treatments via Virtual Stenting 101
4.1 Introduction 101
4.2 Methodology 103
4.2.1 Boundary Conditions 107
4.3 Results 109
4.3.1 Velocity and Flow Rates 109
4.3.2 Pressure 111
4.3.3 Kinetic Energy 113
4.3.4 Wall Shear Stress 113
4.4 Discussion 117
4.5 Mesh Sensitivity 120
4.6 Conclusion 123
References 124

5 Role of Vessel Wall Motion in Aortic Dissection 127
5.1 Introduction 127
5.2 Methods 129
5.2.1 Geometry 129
5.2.2 Boundary Conditions 130
5.3 Results 132
5.3.1 Wall Displacement 132

		5.3.2	Cross-Sectional Area	132
		5.3.3	Wall Stress	134
		5.3.4	Velocity Distribution	137
		5.3.5	Flow Distribution	137
		5.3.6	Pressure Distribution	139
		5.3.7	Proximal and Distal False Lumen	139
		5.3.8	Wall Shear Stress	141
	5.4	Mesh Sensitivity and Efficiency		143
	5.5	Discussion		147
	5.6	Conclusions		150
	References			150
6	**Conclusions and Future Work**			155
	6.1	Introduction		155
	6.2	Main Contributions		156
	6.3	Summary of Main Findings		156
	6.4	Significance of this Study		159
	6.5	Future Work		159

Appendix A: Sample Mesh Images 161

Appendix B: Detailed Mesh Sensitivity Analysis for Virtual-Stenting Simulations 165

Nomenclature

Greek Symbols

α	Womersley number
β	Simplifying term in Windkessel model derivation
γ	Gaussian standard deviation for smooth filter (pixel)
ΔKE	Kinetic energy loss per cardiac cycle (mJ)
Δt	Size of timestep (s)
ϵ	Strain (–)
ε	Turbulence dissipation rate ($m^2 s^{-3}$)
λ	Stretch ratio (–)
λ_{CY}	Carreau–Yasuda time constant(s)
μ	Dynamic viscosity (Pa s)
μ_0	Carreau–Yasuda zero shear viscosity (Pa s)
μ_∞	Carreau–Yasuda infinite shear viscosity (Pa s)
ω	Turbulence frequency (Hz)
ϕ	Arbitrary advected property
ψ	Scaling term in advection solver in ANSYS
ρ	Density (kgm^{-3})
σ	Stress (Pa)
σ_{VM}	Von Mises stress (Pa)
$\sigma_1, \sigma_2, \sigma_3$	Principal stresses (Pa)
τ	Wall shear stress (Pa)
τ_t	Time constant for an RC circuit (s)

Roman Symbols

A_d	Cross-sectional area of deformed geometry (m^2)
A_u	Cross-sectional area of undeformed geometry (m^2)
A^*	Lumina area ratio

Nomenclature

A	Parameter in hyperelastic model
a	Yasuda exponent (a)
B	Parameter in hyperelastic model
C	Compliance (ml/mmHg) or Electrical (F – Farad)
D	Diameter of a tube (m)
D_{lumen}	Vessel wall diameter (m)
E	Young's modulus (Pa)
f	Frequency of cardiac cycle (Hz)
h	Vessel wall thickness (mm)
i	Current (Amps)
$\dot{KE}_i$	Rate of kinetic energy loss (mJ/s)
k	Turbulent kinetic energy ($m^2 s^{-2}$)
L	Length of a tube (m)
M	Mismatch between estimated and experimental parameters
N	Total number of timesteps in a cardiac cycle
m	Carreau-Yasuda Power Law Index (m)
n	Timestep index
P	Pressure (mmHg)
P_{comp}	Computed pressure, either with CFD or LabVIEW (mmHg)
P_{FL}	Pressure in false lumen (mmHg)
P_{clin}	Invasively measured pressure (mmHg)
P_{TL}	Pressure in true lumen (mmHg)
Q	Flow rate (ml/s)
Q^*_{RMS}	Root-mean-square difference in flow rate for Windkessel parameter sensitivity analysis (ml/s)
Q_o	Flow rate using original Windkessel parameters (ml/s)
Q_s	Flow rate using modified Windkessel parameters (ml/s)
R	Resistance: Hydrodynamic (mmhg s/ml) or Electrical (Ω – Ohm)
R_1	Characteristic impedance in Windkessel model (mmhg s/ml)
R_2	Hydrodynamic resistance in Windkessel model (mmhg s/ml)
Re	Reynolds number
Re_c	Critical Reynolds number
St	Strouhal number
T	Length of cardiac cycle (s)
Tu	Turbulence intensity (%)
t	Time relative to start of cardiac cycle (s)
t^*	Time t Normalised by length of cardiac cycle
U	Velocity vector (ms^{-1})
$\overline{U}$	Mean velocity component vector (ms^{-1})
$\tilde{U}$	Periodic velocity component vector (ms^{-1})
U'	Fluctuating velocity component vector (ms^{-1})
V	Voltage (V)

Abbreviations

0D	Zero-dimensional
1D	One-dimensional
2D	Two-dimensional
3D	Three-dimensional
4D	Four-dimensional
AA	Ascending Aorta
AD	Aortic Dissection
BC	Boundary Condition
BT	Brachiocephalic Trunk
CFD	Computational Fluid Dynamics
COPD	Chronic Obstructive Pulmonary Disease
CT	Computed Tomography
CVD	Cardiovascular Disease
DA	Distal Abdominal
DAA	Distal Aortic Arch
FEM	Finite Element Modelling
FSI	Fluid–Structure Interaction
IF	Intimal Flap
LCC	Left Common Carotid Artery
LS	Left Subclavian Artery
MA	Mid-abdominal Aorta
MRI	Magnetic Resonance Imaging
OSI	Oscillatory Shear Index
pcMRI	Phase-contrast Magnetic Resonance Imaging
RRT	Relative Residence Time
TAWSS	Time-Average Wall Shear Stress (Pa)
TEVAR	Thoracic Endovascular Aortic Repair
WK2	Two-element Windkessel Model
WK3	Three-element Windkessel Model
WSS	Wall Shear Stress

List of Figures

Fig. 1.1	Idealised morphologies of the supraaortic branches. **a** Common morphology in which each branch has an independent trunk origin, **b** a bovine aorta, in which the BT and LCC have a common trunk origin	4
Fig. 1.2	Schematic of cleavage formation. A tear forms in the intima layer of the vessel wall and blood moves from the lumen into the region between the intima and media	5
Fig. 1.3	Schematic showing **a** non-communicating and **b** communicating dissections. Entry and re-entry tears are indicated and *grey lines* show flow paths	6
Fig. 1.4	Classification of aortic dissection according to the two main systems. **a** Both ascending and descending aorta are dissected, **b** only ascending aorta is dissected, **c** only descending aorta is dissected............................	7
Fig. 1.5	Illustrations of TEVAR treatment. **a** Typical type-B aortic dissection (*left panel*). The effect of stent-graft placement covering the entry tear is to induce FL thrombosis (*right panel*). **b** Case study showing the additional surgical procedures, including a bare stent and covered stent-grafts in compromised arteries, required to achieve complete FL thrombosis and subsequent remodelling	13
Fig. 2.1	Mesh discretisation in CFX (ANSYS 2011). **a** Control volume definition, **b** Integration points on element edges, **c** tri-linear interpolation for a tetrahedral element......	41
Fig. 2.2	CT scan slices showing **a** the entire domain prior to the cropping stage and **b** the cropped domain in order to reduce the image processing time (only one slice is shown for brevity)	45

Fig. 2.3	CT scan slices showing **a** a representative image, **b** the same image with the mask superimposed mask in *red*	45
Fig. 2.4	CT scan slices with different thresholds. **a** Dissected aorta and the hard tissues with intensity values above 220 shown in *red*, **b** the hard tissues with intensity values above 250 shown in *green*, **c** both masks on the same image, **d** mask resulting from boolean operation subtracting the hard tissues	46
Fig. 2.5	Stages in elimination of the hard tissues. 3D masks of the **a** soft and hard tissue **b** hard tissue only **c** soft tissue only, **d** the smoothed 3D domain	47
Fig. 2.6	The final 3D geometry prior to the cropping stage. **a** The aortic arch, TL and the iliac bifurcation in two views, **b** the *right* anterior and *left* posterior views of the FL	48
Fig. 2.7	**a** Patient-specific reconstructed geometry. **b** The final 3D domain with cropped boundaries (*right*). Selected planes show the TL (*blue*) and FL (*red*) along the descending aorta	49
Fig. 2.8	Surface area at each of the domain boundaries and the largest section of the aneurysm	50
Fig. 2.9	The current entering any node in a circuit is equal to the summation of currents exiting that node	53
Fig. 2.10	Circuit for the two element Windkessel model	54
Fig. 2.11	Circuit for the three element Windkessel model	54
Fig. 2.12	Pressure traces at various locations simulated using 0D modelling (Korakianitis and Shi 2006)	58
Fig. 2.13	Comparison of zero-pressure (*dashed lines*) and Windkessel boundary conditions (*solid lines*). **a** Flow waves and **b** pressure waves at the domain boundaries	59
Fig. 2.14	Pressure for each of the domain outlets relative to the pressure at the AA for zero-pressure (*dashed lines*) and Windkessel (*solid lines*) boundary conditions	60
Fig. 2.15	Comparison of flow-split (*dashed lines*) and Windkessel boundary conditions (*solid lines*). **a** Flow waves and **b** pressure waves at the domain boundaries	62
Fig. 2.16	Pressure for each of the domain outlets relative to the pressure at the AA for flow-split (*dashed lines*) and Windkessel (*solid lines*) boundary conditions	62
Fig. 2.17	Comparison of uniaxial stress-strain behaviour with different material models	64
Fig. 3.1	Block diagram showing the approach used to tune the Windkessel parameters	73

Fig. 3.2	**a** Flow rate at each of the domain boundaries over a single cardiac cycle. **b** Proportion of the flow entering the TL and FL. Vertical *dashed lines* indicate peak systolic flow phase...	75
Fig. 3.3	3D velocity contours at **a** mid-systolic pressure phase, **b** peak-systolic pressure phase, **c** dicrotic notch and **d** mid-diastolic pressure phase	77
Fig. 3.4	Streamlines at **a** mid-systolic pressure phase, **b** peak-systolic pressure phase, **c** dicrotic notch and **d** mid-diastolic pressure phase	78
Fig. 3.5	Streamlines in the aneurysm at peak systolic flow (0.12 s) **a** *left* posterior view and **b** *right* anterior view.......	78
Fig. 3.6	Pressure waves at the domain boundaries.................	79
Fig. 3.7	Pressure contours at **a** mid-systolic pressure phase, **b** peak-systolic pressure phase, **c** dicrotic notch and **d** mid-diastolic pressure phase	80
Fig. 3.8	Wall shear stress distributions in the dissected aorta, *left* posterior and *right* anterior views. **a** TAWSS, **b** OSI	82
Fig. 3.9	Flow characteristics around the tears at two points in the cardiac cycle. Streamlines at (**a**, **e**) peak systole, (**b**, **f**) mid diastole. **c**, **g** Time-averaged WSS vectors to help relate the OSI to the preferential flow direction (**d**, **h**) OSI. *Top* row shows the entry tear, *bottom* row shows the re-entry tear	83
Fig. 3.10	Flow waveforms for **a** +1 mmHg, **b** −1 mmHg. Difference in flow rate between the final RCR parameters and **c** +1 mmHg, **d** −1 mmHg	88
Fig. 3.11	**a** Flow at the AA for all three meshes. **b** Pressure at the AA for all three meshes	90
Fig. 3.12	**a** Flow at the BT for all three meshes. **b** Pressure at the BT for all three meshes..........................	90
Fig. 3.13	**a** Flow at the LCC for all three meshes. **b** Pressure at the LCC for all three meshes	91
Fig. 3.14	**a** Flow at the LS for all three meshes. **b** Pressure at the LS for all three meshes..........................	91
Fig. 3.15	**a** Flow at the DA for all three meshes. **b** Pressure at the DA for all three meshes	92
Fig. 3.16	Coarse mesh: Wall shear stress distributions in the aorta showing *left* posterior and *right* anterior views. **a** TAWSS, **b** OSI..................................	92
Fig. 3.17	Medium mesh: Wall shear stress distributions in the aorta showing *left* posterior and *right* anterior views. **a** TAWSS, **b** OSI..................................	93

Fig. 3.18	Fine mesh: Wall shear stress distributions in the aorta showing *left* posterior and *right* anterior views. **a** TAWSS, **b** OSI..............................	93
Fig. 3.19	Velocity differences between the medium and fine meshes at **a** peak systole and **b** dicrotic notch. Velocity differences between the coarse and medium meshes at **c** peak systole and **d** dicrotic notch......................	94
Fig. 3.20	Pressure differences between the medium and fine meshes at **a** peak systole and **b** dicrotic notch. Pressure differences between the coarse and medium meshes at **c** peak systole and **d** dicrotic notch......................	95
Fig. 3.21	At peak systole: **a** Velocity for the laminar flow model, **b** velocity for the turbulent model, **c** magnitude of the turbulent velocity fluctuations, **d** turbulence intensity.........	96
Fig. 4.1	CT scan slices showing **a** prior to the image segmentation, **b** showing selection of the regions of interest (*red*) and the aneurysm (*blue*), **c** removal of the aneurysm...	104
Fig. 4.2	Geometry of a patient-specific dissected aorta after removing the aneurysm. Inset; *upper panel* shows CT slice at the *dashed line*; *lower panel* shows CT slice, highlighting the ascending aorta and TL in *blue* and the FL in *green*...	105
Fig. 4.3	Patient-specific single-stented reconstructed geometry. **a** Pre-virtual stenting, **b** post-virtual stenting. *Blue lines* correspond to the location of slices selected...............	106
Fig. 4.4	Patient-specific double stented reconstructed geometry. **a** Pre-virtual stenting, **b** post-virtual stenting. *Blue lines* correspond to the location of slices selected...............	107
Fig. 4.5	Pre-operative, single-stent (entry tear occluded) and double-stent (both tears occluded) geometries. The *dashed boxes* indicate the entry tear and re-entry tear in the pre-operative case.................................	108
Fig. 4.6	Flow waveforms at the inlet and outlets of all three cases for one cardiac cycle.............................	109
Fig. 4.7	Velocity magnitude contours at peak systolic flow for all three cases.......................................	110
Fig. 4.8	Streamlines at peak systolic flow for all three cases.........	111
Fig. 4.9	Pressure contours at peak systole for all three cases.........	112
Fig. 4.10	Comparison of pressure waves at the boundaries for pre- and post-operative cases........................	112
Fig. 4.11	TAWSS distribution for all three cases..................	114

Fig. 4.12	Histograms of TAWSS. *Dashed vertical lines* show 5th and 95th percentiles, *solid vertical line* shows median. **a** Pre-operative, **b** single-stent, **c** double-stent.............	115
Fig. 4.13	OSI distribution for all three cases (*right*).................	116
Fig. 4.14	Histograms of OSI. Dashed vertical lines show 5th and 95th percentiles, *solid vertical line* shows median. **a** Pre-operative, **b** single-stent, **c** double-stent.............	117
Fig. 4.15	Histograms of the product TAWSS × (0.5-OSI). **a** Pre-operative, **b** single-stent, **c** double-stent.............	119
Fig. 4.16	Histograms of TAWSS for the mesh sensitivity analysis. *Dashed vertical lines* show 5th and 95th percentiles, *solid vertical line* shows median. *Red* coarse mesh, *Blue* medium mesh, *Green* fine mesh. **a, d, g** Pre-operative, **b, e, h** single-stent, **c, f, i** double-stent	121
Fig. 4.17	Histograms of OSI for the mesh sensitivity analysis. *Dashed vertical lines* show 5th and 95th percentiles, *solid vertical line* shows median. *Red* coarse mesh, *Blue* medium mesh, *Green* fine mesh. **a, d, g** Pre-operative, **b, e, h** single-stent, **c, f, i** double-stent	122
Fig. 5.1	Geometry of **a** Solid **b** fluid. Entry and re-entry tears are indicated, as is the co-ordinate z. The domain boundaries are AA ascending aorta; DA distal abdominal; LS *left* subclavian artery; LCC *left* common carotid; BT brachiocephalic trunk. *Dashed green* boxes show the regions analysed in Figs. 5.9 and 5.10............	130
Fig. 5.2	Displacement of the vessel wall at various time instances. Contours show the displacement relative to the undeformed geometry in the *left* posterior view at **a** mid systole, **b** peak systole and in the *right* anterior view, **c** peak systole and **d** dicrotic notch................	133
Fig. 5.3	Cross sectional area of each lumen and the combined area along the descending aorta for the rigid wall model, A_u. See Fig. 5.1 for co-ordinate system..................	134
Fig. 5.4	Map of lumina area ratio $A^* = (A_d - A_u)/A_u \times 100\%$, against t^* (relative time in cardiac cycle) and z. **a** True lumen, **b** false lumen	135
Fig. 5.5	Von Mises stresses in the vessel wall in the *left* posterior view at **a** mid systole, **b** peak systole and in the *right* anterior view, **c** peak systole and **d** dicrotic notch	136
Fig. 5.6	Streamlines in the *left* posterior view at **a** mid systole, **b** peak systole and in the *right* anterior view, **c** peak systole and **d** dicrotic notch	137
Fig. 5.7	Flow distributions in the descending aorta. Each subfigure compares the flow rate across the indicated	

	plane for FSI and rigid wall simulations against t^*. *Red*—true lumen, *blue*—false lumen. Vertical *green* lines in Proximal FL and Distal FL subfigures indicate time instances considered in Figs. 5.9 and 5.10	138
Fig. 5.8	Pressure contours in the *left* posterior view at **a** mid systole, **b** peak systole and in the *right* anterior view, **c** peak systole and **d** dicrotic notch	140
Fig. 5.9	Flow characteristics in the proximal FL. The *left* column shows streamlines calculated from the Proximal plane in the FL (see Fig. 5.7). Each row shows a time instance as indicated by the insets and the vertical *green* lines in Fig. 5.7. The *right* column shows pressure contours relative to the pressure at the AA (Pressure minus Pressure at AA)	141
Fig. 5.10	Flow characteristics in the distal FL. The *left* column shows streamlines calculated from the Distal plane in the FL (see Fig. 5.7). Each row shows a time instance as indicated by the insets and the vertical *green* lines in Fig. 5.7. The *right* column shows pressure contours relative to the pressure at the DA (Pressure minus Pressure at DA)	142
Fig. 5.11	Wall shear stress characteristics. **a** TAWSS distributions for the FSI simulation, **b** dynamic viscosity distribution at peak systole, **c** percentage difference in TAWSS relative to the rigid wall model	142
Fig. 5.12	OSI characteristics. **a** OSI distributions for the FSI simulation, **b** percentage difference in OSI relative to the rigid wall model	144
Fig. 5.13	Influence of the mesh refinement on the boundary conditions. *Solid lines* medium mesh, *dashed lines* fine mesh. **a** Pressure at the boundaries, **b** Flow rates at the boundaries	145
Fig. 5.14	Displacement at peak systole for **a** the medium mesh and **b** the fine mesh. **c** Difference in displacement between the two meshes	146
Fig. 5.15	Wall shear stress indices for the two meshes. TAWSS for **a** medium and **b** fine meshes. OSI for **c** medium and **d** fine meshes	147

List of Tables

Table 3.1	Minimum and maximum pressure values at various locations throughout the domain, measured clinically and derived from the simulations (discussed later in this section). All pressures in mmHg. Difference rows show the difference between the measured and simulated values as a percentage of the AA pulse pressure	72
Table 3.2	Final tuned Windkessel parameters	74
Table 3.3	Summary of the Windkessel parameters calculated in the sensitivity analysis. See Eq. 3.6 for definition of Q^*_{RMS}	88
Table 4.1	Maximum differences between the pressure and flow estimates at the boundaries obtained with different meshes	120
Table 5.1	Parameters used for Carreau-Yasuda blood viscosity model (Gijsen et al. 1999) and hyperelastic wall model of (Raghavan and Vorp 2000)	131

Chapter 1
Introduction

This chapter will begin with an overview of the motivation for the research carried out in this thesis on the numerical modelling of aortic dissection (AD). Subsequently, an introduction to relevant cardiovascular physiology and pathology will be provided, followed by a review of modelling approaches for cardiovascular disease and more specifically AD. This chapter will conclude with the identification of a number of outstanding issues that will be addressed in this thesis.

1.1 Motivation and Background

Cardiovascular disease (CVD) is the leading cause of death in the world (Yacoub et al. 2014) and in most of Europe CVD accounts for around 40% of all mortality (Kromhout 2001). Despite significant improvements in mortality rates for CVD overall (Svensson et al. 1990), AD remains an extremely severe condition with high mortality rates (Delsart et al. 2013). AD is one of the most catastrophic cardiovascular events (Sbarzaglia et al. 2006) and although the incidence of the disease is relatively rare (5–35 per million per year (Khan and Nair 2002; Meszaros et al. 2000)), the outcomes are dire, with 20% of patients dying before reaching the hospital and a further 30% dying during admission (Masuda et al. 1991). For those patients that survive, the choice of treatment is critical and 70% of all deaths caused by AD occur within the first 2 weeks (Crawford 1990). Additionally, in one study a failure to accurately diagnose AD was observed in 28% (Spittell et al. 1993) of patients. Although modern advances in imaging modalities have improved diagnosis, the disease remains difficult to diagnose and treat (Erbel et al. 2001).

For cases of AD involving the ascending aortic arch ('Stanford type-A'), immediate surgical repair is generally required. However, the treatment for dissections not involving the aortic arch ('Stanford type-B') is more variable and patient-specific. In such cases, the relative risks of withholding surgical treatment compared to surgery, and the choice of surgical approach, are less clear. Although successful surgical intervention will have a favourable outcome for the patient, due to the complexity of the procedure and the danger of influencing arteries branching from the aorta, there is a high level of risk associated with carrying out the surgical procedure along the dissected abdominal aorta (Khan and Nair 2002), as well as a corresponding risk of post-operative complications such as respiratory failure, renal failure and paraplegia (Sbarzaglia et al. 2006). Unpredictable outcomes, such as complete blockage of the TL by expansion of the thrombosed FL (Takahashi et al. 2008), or thrombi occluding the intercostal arteries (Yamashiro et al. 2003), can lead to paraplegia, defined as weakness or complete loss of function of the lower limbs (Bozinovski and Coselli 2008). Where the initial presentation is not severe, the first treatment approach for patients presenting with type-B AD is to treat pharmacologically, with the aim of alleviating symptoms by decreasing and stabilising blood pressure and heart rate (JCS Joint Working Group. 2013). Patients are monitored, and regular check-ups are advised. For patients arriving in hospital in a critical condition, or if complications such as aortic expansion, malperfusion or aneurysm formation arise in patients undergoing pharmacological treatment, surgical intervention must be considered (Erbel et al. 2001; Roberts 1981). This decision is made on a patient-to-patient basis by weighing up benefits of successful surgery against risks of confounding the disease (Umaña et al. 2002).

In such situations, it is essential to have as much information as possible on the potential outcomes for a given patient. This need (for cardiovascular disease in general) has led to the development of 'patient-tailored' or 'patient-specific' numerical modelling to provide information to clinicians on haemodynamic parameters (wall shear stresses, pressures etc.) (Taylor et al. 1999). Ensuring that the results of such modelling are sufficiently accurate, whilst still being feasible for use in clinical settings, is an essential requirement if these models are to be used for clinical diagnosis and/or treatment planning. Critically, application of appropriate boundary conditions (BCs) and simplifying assumptions dictate the utility of the results. In the context of AD, wherein morphological (geometry of the aorta), structural (vessel wall structure) and fluid (altered pressures and flows) aspects may all be substantially modified, modelling complexity and computational expense are increased.

The aim of the present thesis is to address a number of outstanding issues regarding the simulation of haemodynamics in AD, as part of the development of a hollistic computational framework to investigate this pathology *in silico*.

1.1.1 Introduction to the Circulatory System

Aortic flow and pressure are predominantly the outcome of the interaction between the heart and the arterial system. The heart pumps the oxygenated blood into the systemic vasculature in a rhythmic cycle. The systemic vessels carry and distribute the blood to the different organs, collect the deoxygenated blood and direct it back towards the heart and lungs. The vessels vary in their functionality, mechanical

1.1 Motivation and Background

properties and size. Any deviation from physiological conditions within the cardiovascular system may lead to irreversible and life threatening outcomes (Drake et al. 2010).

The vasculature consists of a closed-loop fluid system, with pressure provided by the heart. The heart is situated between the lungs with two thirds of the mass on the left side of the body (Tortora and Grabowski 2001). The heart consists of four chambers; the two upper ones are the atria and the lower ones are the ventricles. The two atria are divided by a very thin layer called inter-atrial septum, and the ventricles by the inter-ventricular septum. The deoxygenated blood enters the right atrium through the vena cava. Blood then enters the right ventricle through the tricuspid valve and is pumped to the pulmonary artery, which carries the blood to the lungs, wherein the blood is oxygenated. The oxygenated blood comes back from the lungs and enters the left atrium and then flows through the mitral valve to the left ventricle. As the heart contracts, the pressure gradient increases between the ventricle and the aorta, across the aortic valve, until it reaches some threshold value, at which point the valve opens and blood is ejected into the ascending aorta. As the muscle of the left ventricle relaxes, the blood in the aorta tends to flow back towards the heart and in this way closes the aortic valve. The contraction and relaxation of the heart forms the cardiac cycle, with systolic and diastolic phases respectively. The volume of blood exiting the left ventricle per minute is called cardiac output and the amount of blood ejected at each contraction is known as the stroke volume (Tortora and Grabowski 2001; Drake et al. 2010).

Having left the heart, blood enters the aorta, which is the largest artery in the vasculature. The most common morphology of the aorta is an aortic arch with three branch points along the upper wall, as shown in Fig. 1.1a. The first branch (the one closest to the heart) is the brachiocephalic trunk (BT), which is the largest of the three branches. The second branch after the BT is the left common carotid artery (LCC), followed by the left subclavian artery (LS). This sequence is found in 65–80% of people (Mligiliche and Isaac 2009). However, other anatomical variations have been observed in terms of the number and order of the branches, such as a 'bovine arch' (Fig. 1.1b). Bovine arch is a misnomer, which incorrectly refers to the common trunk origin for both BT and LCC "resembling that of a cow" (Layton et al. 2006) (which incidentally it does not). People with a bovine arch have an increased risk of thoracic aorta dilation and ascending aorta aneurysm formation, which may require surgical intervention (Malone et al. 2012).

Arteries are hollow tubes through which the blood flows and as they sequentially bifurcate towards the arterioles, their cross sectional area reduces. They deliver oxygenated blood to the microcirculation, which consists of arterioles, capillaries and venules. The veins return the blood to the lungs for re-oxygenation. The arterial wall consists of three layers. The inner layer is called the intima and is composed of the endothelium, basement membrane and elastic lamina (Tortora and Grabowski 2001; Drake et al. 2010). The middle layer is called the media and consists of smooth muscle and elastic tissues. The adventitia is the outer layer of the aorta and consists of predominantly elastic and collagen fibres. The elastic nature of these fibres means that they expand during systole, and slowly contract during diastole, ensuring a

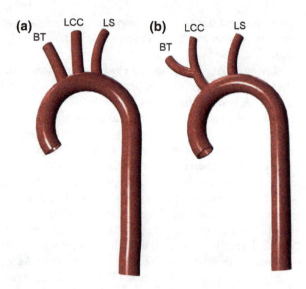

Fig. 1.1 Idealised morphologies of the supraaortic branches. **a** Common morphology in which each branch has an independent trunk origin, **b** a bovine aorta, in which the BT and LCC have a common trunk origin

continuous supply of blood to the downstream organs and tissues. This function was compared to the Windkessel (literally 'air chamber') used to pump old German fire hoses to smooth out the pumping action (Shi et al. 2011), and is a key component of 0D (zero-dimensional) or 'Windkessel' modelling, as will be discussed in detail later.

As the arteries reduce in diameter below around 0.1–0.3 mm, they are termed arterioles. The arterioles have similar structure to arteries, but with differences in the proportion of each of the wall layers. Their role is particularly important in the circulatory system, with 50–60% of the total pressure drop occurring in these vessels (Fronek and Zweifach 1975). Furthermore, the smallest arterioles supply flow to the capillary bed, and the arterioles act to regulate capillary pressure via vasoconstriction and vasodilation (Tortora and Grabowski 2001; Drake et al. 2010).

The capillaries are microscopic vessels, which have considerably thinner walls than the arterioles. They act as a bridge that connects arterioles to venules, and within the capillary bed, the majority of nutrient and waste exchange, as well oxygen delivery occurs. The number of capillaries around an organ or tissue is dependent on the organs metabolic requirement, i.e. kidneys and livers have more capillaries than ligaments or tendons (Tortora and Grabowski 2001).

Venules have a smaller proportion of smooth muscle than the arterioles. Venular walls close to the capillaries are thin and increase in thickness with distance from the capillaries. Veins are similar to the arteries but have thinner walls and significantly less smooth muscle (Tortora and Grabowski 2001). There is minimal pressure in the venous side of the circulation, and hence veins have valves to restrict flow reversal. However, this is not sufficient to maintain and direct blood towards the heart and the skeletal muscle pumps and the respiratory pump assist the veins in returning the blood flow to the heart (Drake et al. 2010).

1.1.2 Aortic Dissection

Aortic diseases can be inherited, inflammatory, traumatic or degenerative (Erbel et al. 2001). Thoracic aortic diseases occur at a rate of 16.3 and 9.4 cases per 100000 per year for men and women respectively in Europe (Bastien et al. 2012). The range of aortic diseases is wide, including coronary and peripheral artery disease, atherosclerosis, intramural haematoma, aortic aneurysm and aortic dissection (Erbel et al. 2014).

1.1.2.1 Description of the Disease

AD was first described over 200 years ago (Hagan et al. 2000). In AD, a tear forms in the intimal layer of the vessel wall, and blood forces the vessel wall layers apart. This condition can be initiated either by cleavage formation (see Fig. 1.2) in the intimal layer of the vessel wall (Erbel et al. 2001) or by an intramural haematoma (blood pooling between the vessel wall layers), which subsequently causes the intimal layer to tear (Wilson and Hutchins 1982). Under arterial pressure, blood enters the vessel wall and propagates in both longitudinal and circumferential directions, creating an

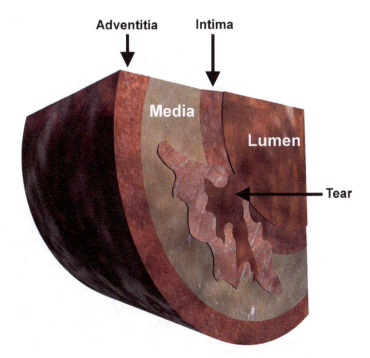

Fig. 1.2 Schematic of cleavage formation. A tear forms in the intima layer of the vessel wall and blood moves from the lumen into the region between the intima and media

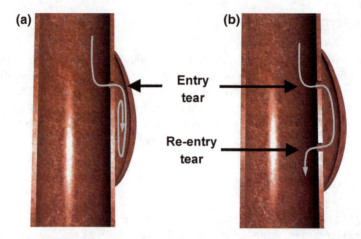

Fig. 1.3 Schematic showing **a** non-communicating and **b** communicating dissections. Entry and re-entry tears are indicated and *grey lines* show flow paths

additional flow channel known as the false lumen (FL) (Khan and Nair 2002; Erbel et al. 2001; Roberts 1981). Sudden formation of an intimal lesion can occur due to mechanical stresses acting on the elastin components (Nguyen et al. 2002), an increase in flexion stress, or excessive movement of the aorta (Khan and Nair 2002; Khanafer and Berguer 2009). Khanafer and Berguer (2009) found that the difference in elasticity between the aortic wall layers has a significant impact on the generation of AD and identified higher wall stresses in the media layer.

AD is a rare disease, and its occurrence rate is approximately 0.5–1% compared to that of coronary artery disease (CAD). An occurrence rate for AD of 5–35 cases per million per year has been reported (Khan and Nair 2002; Meszaros et al. 2000). However, this may be an underestimate of the true prevalence as many patients, particularly the elderly, are not autopsied and thus the condition is not discerned from general cardiovascular failure. In large series of such patients who were autopsied, a prevalence of 0.2–0.8% was observed (Tsai et al. 2009; Levinson et al. 1950).

ADs can be communicating or non-communicating as shown schematically in Fig. 1.3. AD is classified as non-communicating (Fig. 1.3a) when there is no second tear and hence the blood in the FL is almost stationary. Non-communicating AD can occur due to the diseased vessel wall or haematoma (Erbel et al. 2001; Roberts 1981). In non-communicating AD, the stagnant flow in the FL can lead to thrombosis. Communicating AD (Fig. 1.3b), occurs due to the formation of a second tear, which leads to blood flow through the FL, parallel to the main blood stream or 'true lumen' (TL). Tsai et al. (2007) found that patients with communicating AD had the lowest mortality rates, followed by those with complete thrombosis. Patients with only partial thrombosis (classified based on the presence of both thrombosis and flow) had the highest mortality rate (32%).

1.1 Motivation and Background

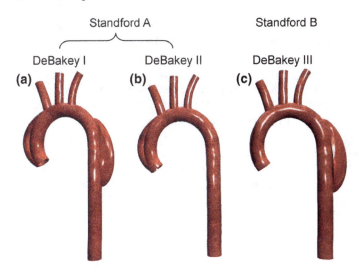

Fig. 1.4 Classification of aortic dissection according to the two main systems. **a** Both ascending and descending aorta are dissected, **b** only ascending aorta is dissected, **c** only descending aorta is dissected

AD is classified according to the anatomical location of the tear and FL. The Stanford classification system distinguishes between type-A (any dissection involving the ascending aorta—Fig. 1.4a, b) and type-B (dissections occurring elsewhere—Fig. 1.4c). An alternative classification system is named after DeBakey, who was the first person to surgically repair an AD (De Bakey et al. 1955). This classification subdivides the onset and the progression of the dissected aorta in a more specific way; 'type I' when both ascending and descending aorta are involved, 'type II' when dissection occurs only in the ascending aorta and 'type III' when dissection occurs only in descending aorta (Fig. 1.4). Note that DeBakey type I and II would be classified as Stanford type-A, and DeBakey type III is equivalent to Stanford type-B. Svensson et al. (1999) proposed a more detailed 5-class system, involving such factors as presence or absence of clots, plaques and some causative factors. Class 1 involves the classic formation of AD through separation of the intima from the media causing two distinct lumina. Class 2 covers dissections occurring from intramural haematoma with no visible flap or tear. Class 3 describes an intimal tear without haematoma and an eccentric ascending aortic bulge. Class 4 defines (generally subadventitial) atherosclerotic ulcers and Class 5 refers to dissection arising from iatrogenic trauma. However, the Stanford classification system remains the most commonly used, and will be adopted herein.

Type-A dissections account for 50–60% of all dissections (Tsai et al. 2009; Bakey et al. 1965) and require open surgery in the majority of cases (Capoccia and Riambau 2014). For type-B dissections, the presence or absence of complications is a significant factor in the prognosis. For type-B dissections with no complications, the survival rate is 80–90% for the first year (Rango and Estrera 2011) and 50–80% over five years (Estrera et al. 2006; Hagan et al. 2000). In such cases, the condition is generally managed medically (Erbel et al. 2001). Whether pre-emptively performing endovascular treatment, in addition to medical management, is favourable is not yet clear. A two-year follow up study showed no significant difference in mortality between patients who underwent endovascular treatment, although longer term prognoses may differ (Nienaber 2011). In type-B dissections with complications, 14–67% of patients are at risk of permanent spinal damage or post-operative mortality (Svensson et al. 1990; Glower et al. 1991). Paraplegia can occur if several intercostal arteries are blocked by the IF or separated from the aortic lumen (Erbel et al. 2001; Weigang et al. 2008a). Endovascular surgery appears to be favourable to open surgery (Fattori et al. 2008b), but the long term outcomes have not yet been established.

Temporally, AD has generally been classified as acute for the first two weeks, and chronic thereafter (Criado, 2011). This time period is related to the fact that 70% of deaths in AD occur within the first 14 days after onset of the disease (Crawford 1990). An additional category termed 'sub-acute' was introduced (Nienaber et al. 2009; Svensson et al. 2008), before the international registry for aortic dissection (IRAD) defined a new classification system (Booher et al. 2013). The first 24 hours are classified as 'hyper-acute', follow by acute to the end of week one, sub-acute up to one month, and chronic thereafter. This improved classification system provides clinicians with better guidelines from which to make decisions regarding treatment options.

Nonetheless, a recent prospective analysis of the literature on treatment of type-B AD, came to the the conclusion that "literature results were largely heterogeneous and should be interpreted cautiously" (Fattori et al. 2013). Thus, treatment options for type-B AD are not well defined and the choice of surgical approach retains a subjective element (Capoccia and Riambau 2014). This may be due to the fact that the range of different 'complications' associated with type-B is vast, possibly precluding the reductionist approach required in generalising treatment plans.

1.1.2.2 Outcomes of Aortic Dissection

Mortality: The mortality rate varies depending on the type of AD. For untreated patients with type-A dissection, it increases by 1–3% per hour (Erbel et al. 2001; Hirst et al. 1958); with 25% in the first 24 hours, 70% in the first week and 80% in the first two weeks after disease initiation. Only 10% of the patients with proximal AD live for a year and nearly all patients die within the 10 years of the disease onset (Khan and Nair 2002). The indication for surgery or medical treatment has a significant impact, with mortality rates of $\approx$27% for those receiving surgery and $\approx$58 for those who did not (although withholding of surgery was often due to advanced age or the presence of an additional condition) (Hagan et al. 2000; Tsai et al. 2009).

For uncomplicated type-B dissections, the in-hospital survival rate is 90% (Tsai et al. 2009; Svensson et al. 2008) although for patients who only undergo medical treatment, the 5 year survival rate is worse than for type-A patients who have undergone surgery (Patel 1986). Overall survival rates for type-B dissections are 70–85% in the first year (Masuda et al. 1996; Erbel et al. 1993) and 60% after two years (Erbel et al. 1993). The mortality rate of patients with type-B dissection that underwent surgical intervention has been observed to be ≈30% (Szeto et al. 2008; Hagan et al. 2000; Booher et al. 2013). Long term survival rates for patients undergoing medical or endovascular treatment are reported to be 10%.

Rupture: Vessel wall rupture in AD is associated with a very high mortality rate of 62.5% (Szeto et al. 2008; Trimarchi 2006). It is one of the most severe 'complications' in type-B AD, and indications of possible rupture suggest that pre-emptive endovascular treatment may be necessary (Nienaber et al. 2014). Patients presenting with hypotension are likely to have suffered from rupture (Criado, 2011). In a study of 50 patients with chronic type-B dissections, 9 patients died from fatal rupture (Juvonen et al. 1999). Around 60% of late stage deaths are due to rupture, as long term patent FL (free from thrombosis) often leads to aneurysmal dilatation (Nienaber et al. 2014).

Enlargement of the aorta is a common occurrence in type-B dissections (20–40%) (Schor et al. 1996; Juvonen et al. 1999; Gysi et al. 1997), and one of the main risk factors for rupture is the diameter of the aorta. For diameters greater than 55–60 mm, there is a 30% chance of rupture (Fattori et al. 2013). Svensson et al. (2008) report a 34% risk of rupture for ascending aorta diameters greater than 60 mm and a 43 % chance of rupture for descending aorta diameters exceeding 70 mm. False lumen patency has also been indicated as a risk factor for rupture (Bernard et al. 2001), as have signs of haematoma (Nienaber and Eagle 2003).

Monitoring the patient and acquisition of CT images can help clinicians to identify risk of rupture. In order to prevent rupture, the systemic blood pressure should be stabilised and decreased using β-blockers in combination with anti-hypertensives and analgesics (JCS Joint Working Group. 2013; Tsai et al. 2009). In the event of aortic rupture, stenting of the rupture and the entry tear, or in the most severe cases the entire thoracic aorta, is required. The stent must cover the rupture and the primary entry tear site in order to be successful (Szeto et al. 2008). Endovascular treatment appears to be successful at preventing aortic rupture (only 0.8% was observed in a long term thoracic endovascular aortic repair (TEVAR) study (Nienaber 2011; Parker and Golledge 2008)).

Major branch vessel obstruction: Obstruction of the branches sprouting from the descending aorta can lead to a reduction of blood flow into these branches (such as the renal arteries) and malperfusion syndrome. Although previously considered in terms of static and dynamic mechanisms (the former occurs when the pressurised FL shrinks the branches, and the latter is as a result of the intimal flap blocking the blood stream (Criado, 2011)), it has become clear that the TL is occluded by compression from the pressurised FL (Criado, 2011; Yang et al. 2014). Malperfusion leads to serious problems for the organs and tissues downstream of the obstructed blood vessel, causes additional complications and jeopardises the end-organ perfusion (Tsai et al. 2009; Svensson et al. 2008). Patients undergoing surgery for type-B dissection presenting with malperfusion had a 28% in-hospital mortality rate (Trimarchi 2006).

Severe pain and additional progression of FL can also occur as a result of vessel obstruction (Tsai et al. 2009). Rapid decisions are required by clinicians once severe malperfusion is indicated. Malperfusion in type-B AD could affect the bowel, liver and kidneys (Khan and Nair 2002; Criado, 2011). Renal malperfusion syndrome can lead to acute renal failure, along with anuria (suppression of the kidney's ability to form urine), increased blood pressure and eventually renal failure within hours of the complete blood flow obstruction (Fattori et al. 2008a). In lower limb malperfusion syndrome, reduced blood pressure pulsations occur in the femoral or iliac arteries and can the condition can lead to severe pain (Fattori et al. 2008b).

In malperfusion syndrome (organ ischaemia), surgical repairs become vital and vessel adjustment is required to enlarge the obstructed vessel. This is usually done using TEVAR. Moreover, balloon fenestration may also be involved in the process of invasive treatment for type-B sufferers (Erbel et al. 2001; Szeto et al. 2008). Malperfusion treatments vary for static and dynamic cases. For static blockages, the priority is to reperfuse the vessel and in the case of dynamic malperfusion, clinicians aim to fenestrate the flap by creating a large re-entry tear (Patel et al. 2009). However, this is a temporary solution, as it does not allow for vessel remodelling and thus only treats the malperfusion and not the dissected aorta itself (Patel et al. 2009).

1.1.2.3 Risk Factors for Aortic Dissection

Gender and age: AD is very rare in children and adolescents, with around 3.5% of AD sufferers fitting this category (Ngan et al. 2006). Men are more prone to get AD than women, particularly in the younger age groups (< 50 years old), wherein 80% of sufferers are men, compared to 50% of patients over 75 years old (Nienaber 2004). However, unfavourable outcomes are more common for women suffering from AD (Capoccia and Riambau 2014; Nienaber 2004); 63% of AD patients who experience vessel wall rupture are female, as opposed to 37% being male (Hagan et al. 2000; Tsai et al. 2009). Female patients tend to get AD at later stage of life than male patients (Hagan et al. 2000).

Arterial hypertension: Hypertension is the most common factor in AD initiation and studies have reported that it is present in around 62–78% of patients with AD (Khan and Nair 2002). In another study of 464 AD patients, 72% were reported to have a history of arterial hypertension (Hagan et al. 2000). The stress on the wall increases with the pressure for a given vessel thickness and radius. Hence at elevated blood pressures, the wall stresses can become very high, increasing the risk of cleavage formation. The treatment of hypertension is crucial for AD patients, and specialists aim to stabilise systolic blood pressure to 100–120 mmHg (Erbel et al. 2001).

Aortic diseases: A number of other aortic diseases have been observed to be complicit in the generation of AD. These include excessive dilation of the aorta and aortic root, aortic aneurysm, anuloaortic ectasia, aortic arch hypoplasia, coarctation of aorta, aortic arteritis and bicuspid aortic valve (Khan and Nair 2002; Burchell 1955). Up to

30% of AD patients are found to be suffering from prior aortic conditions (Alfonso et al. 1997; Svensson and Crawford 1992; Spittell et al. 1993).

Hereditary connective tissue disorders: A number of connective tissue disorders have been observed to be responsible for AD initiation and progression, thus it is important to know their influence on AD and therapeutic approaches or surgical management used in their treatment.

Marfan syndrome is responsible for 5% of AD patients and mostly affects the ascending part of the aorta (Stanford type-A) (Tsai et al. 2009; Hagan et al. 2000). Marfan syndrome is a hereditary condition that affects the ocular, skeletal and cardiovascular systems (Dietz et al. 1991), with an incidence of 1/5000. A strong correlation has been found between Marfan syndrome and a genetic mutation in the 'fibrilin-1' gene, which is responsible for generating fibrilin; a constituent part of the extracellular matrix (Erbel et al. 2001; Sakai et al. 1986). Patients with Marfan syndrome have abnormal aortic distensibility and stiffness indices (Adams et al. 1995; Pyeritz 2000) (increased stiffness). Correspondingly, the dilation of the aorta is expected to be reduced in the presence of this syndrome.

Ehler-Danlos is an autosomal dominant genetic disorder which causes articular hypermobility, skin hyperextensibility and tissue fragility (Beighton et al. 1998; Braverman 2011). The syndrome is caused by structural defects in the pro $(\alpha\text{-}1)$ (III) chain of collagen type III (Erbel et al. 2001).

Trauma to the aorta: Iatrogenic trauma (that accidentally caused by medical examination or treatment) can also lead to AD. Around 5% of AD cases occur as a result of medical examination or surgical procedures such as aortic valve repair and bicuspid aortic valve repair. In the majority of patients in which AD was due to iatrogenic initiation, the dissection was observed to involve the descending aorta (Khan and Nair 2002; Criado, 2011).

Atherosclerosis: In type-A AD, 31% of patients were reported to have a history of atherosclerosis, although evidence suggests that atherosclerosis is not a risk factor for AD initiation *per se* (Coady et al. 1999). In type-B AD, a history of atherosclerosis is an independent risk factor for post-operative mortality (Tsai et al. 2006). Atherosclerosis is associated with intramural haematoma and atherosclerotic aortic ulcer (classified as variants of AD (Khan and Nair 2002)), and additionally with increased probability of iatrogenic trauma, due to the weakening of the vessel wall structure (Litchford et al. 1976).

Substance abuse: Cocaine use can also lead to AD, due to its effect of rapidly elevating blood pressure (Perron and Gibbs 1997; Fisher and Holroyd 1992). In such patients, the initiation of a Stanford type-A dissection tends to occur within 24 hours of the cocaine consumption, which can be confirmed by laboratory tests. In general, patients with cocaine initiated AD tend to be younger than those with the disease caused by other factors (Daniel et al. 2007).

Pregnancy: Studies have shown that physiological changes during pregnancy could lead to AD (Burchell 1955; Nguyen et al. 2002). A number of case reports describe specific conditions for this relatively rare occurrence (Kinney-Ham et al. 2011; Kohli et al. 2013; Aziz et al. 2011).

1.1.2.4 Treatment of Type-B Aortic Dissection

As intervention is generally necessary in type-A AD, development of additional tools for diagnosis promises fewer benefits than in the case of type-B dissections. Herein, the focus will thus be on type-B dissection. As previously mentioned, the first treatment approach for uncomplicated type-B dissection is medical management (pharmacological), in order to decrease and stabilise the blood pressure. However, medical treatment is not always the best option for type-B sufferers, as additional complications may occur during the different phases of the disease progression, i.e. malperfusion or rupture (Khan and Nair 2002; Erbel et al. 2001; Hagan et al. 2000). Even in the absence of clear complications on initial presentation, the long term (5 year) prognosis for such patients is that 50% will experience aortic complications.

These complications require intervention, however the decision regarding how and when to operate retains a degree of subjectivity (Capoccia and Riambau 2014). In moribund patients, open surgery is often a last resort, and correspondingly has high risk associated with it; 7% of patients become paraplegic, 43% have cardiac complications and the overall mortality rate is 18–50% (Shaw 1955; Trimarchi 2006; Bozinovski and Coselli 2008). Endovascular treatment or TEVAR, which has lower associated risks, is the preferred treatment approach where possible.

The standard approach for TEVAR is to use a stent to cover the entry tear. By closing the entry tear, the FL is isolated from the main blood supply, enhancing likelihood of FL thrombosis (Nienaber et al. 2005). In successful TEVAR, wherein FL thrombosis occurs (Grabenwoger et al. 2012), remodelling takes place, and the TL cross section increases (Nienaber et al. 1999).

Figure 1.5a shows this schematically. In the left panel, the entry and re-entry tears can be seen, allowing communication between the TL and the expanded FL. In this case, the blood flow in the FL would maintain its patency, indicating risk of aneurysmal dilation for the patient (Nienaber et al. 2014). In the right panel, a stent is placed covering the entry tear, and an additional bare stent is added to support the TL below the tear. A thrombus is formed in the FL, which would eventually lead to remodelling. However, patient-specific geometry and haemodynamics influence the extent of FL thrombosis, as the presence of fenestrations due to branches or re-entry tears can allow FL flow to continue, and thus maintain the patency of some portion of the FL (Mossop et al. 2005). Figure 1.5b shows an example of this from a case study (Mossop et al. 2005). In the left panel, a dissected aorta is illustrated with an entry tear at the distal aortic arch, a small entry tear in the subclavian artery, a tear in the abdominal aorta and a fourth small tear in the left common iliac artery. Additionally, the entrance to the left renal artery is obstructed by the IF. TEVAR treatment was used to block the entry tears and a week later a bare stent was added to restrict TL

1.1 Motivation and Background

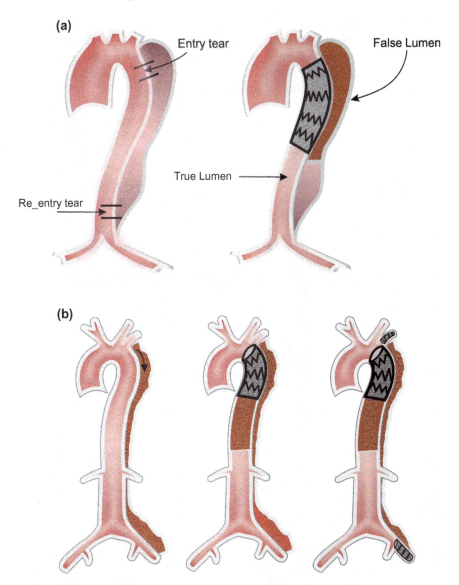

Fig. 1.5 Illustrations of TEVAR treatment. **a** Typical type-B aortic dissection (*left panel*). The effect of stent-graft placement covering the entry tear is to induce FL thrombosis (*right panel*). **b** Case study showing the additional surgical procedures, including a bare stent and covered stent-grafts in compromised arteries, required to achieve complete FL thrombosis and subsequent remodelling

collapse in the thoracic aorta. However, the re-entry tears in the left common iliac and infrarenal aorta persisted and thus the distal FL remained patent. Follow-up surgery using covered stent-grafts in the compromised branches, in combination with coil embolisation successfully induced thrombosis of the distal FL and aortic remodelling was achieved within a year. Whilst the additional interventions for this patient were successful, Scali et al. (2013a) found in their study that only 21% of 63 patients with chronic AD treated with TEVAR had complete thrombosis, with the remaining patients retaining some portion of FL patency. Mani et al. (2012) only observed aortic remodelling in 50% of the 58 patients considered in their study, with the remodelling occurring in cases of extensive thrombosis. De-pressurisation of the false lumen is indicative of reduced flow and thus successful thrombosis. Although measurement of pressure in the FL with catheters is not commonly carried out, presumably due to the fragility of the damaged vessel wall, it is possible to monitor pressure in the FL using wireless CardioMEMS system (Parsa et al. 2010). This device works by changing capacitance and thus natural frequency in response to pressure changes, and can thus be activated and read using radiofrequency excitation (Parsa et al. 2010). Parsa et al. (2011) used this device to investigate the reduction in the FL pulse pressure ratio (ratio of pulse pressure in FL to systemic pulse pressure) after TEVAR in patients with chronic type-B dissection. They observed a change from 52 to 14% on follow up (an average of 27 months). Such technology shows great potential for future wireless monitoring of the efficacy of TEVAR treatments.

Hanna et al. (2014) reported on the 5–7 year outcome of 50 patients undergoing TEVAR surgery. The survival rate was 84% at 5 years and none of the deaths were related to aortic complications. Although one in four patients required reintervention, this data strongly supports the use of TEVAR. Qin et al. (2013) compared TEVAR treatment of complicated type-B AD patients to pure medical management for uncomplicated cases. Adverse events were experienced by 3% of each group in the first year and around 35% after three years. However, after 5 years, only one in three was free from adverse events in the medically managed group, as compared to two in three for the TEVAR group. Nienaber et al. (2013) analysed a randomised trial on 140 patients with uncomplicated dissections (the INSTEAD XL trial), with approximately half receiving pre-emptive TEVAR treatment and all undergoing medical management. The results showed that in the short term (2 years), those undergoing TEVAR had a marginal, but not statistically significant, increase in the rate of all cause mortality and very similar disease progression. However, in the period from 2 to 5 years, of those who had received TEVAR treatment, there were no mortalities and very few complications, as compared to the medical only cohort, who had an approximately 80% survival rate. The authors concluded that 'uncomplicated dissection', meaning initially stable, may be a misnomer and that pre-emptive TEVAR treatment would benefit suitable patients in the long-term. Other reports were not so favourable for TEVAR. Jones et al. (2014) reported statistically significant but small (3–7%) increased rates of chronic obstructive pulmonary disease (COPD), congestive heart failure, diabetes and renal failure, as compared to open surgery. TEVAR is a relatively recent development and more time is required before conclusive long-term risk factors can be established (Grabenwoger et al. 2012).

Umaña et al. (2002), stated that individually tailored patient treatment is important when it comes to dissections involving the descending aorta, and suggested as a guideline that surgery should be carried out where a less than 5% chance of operative mortality can be achieved. The need for patient-tailored approaches is due to the highly important effects of the patients geometry, haemodynamic factors such as blood flow, pressure and wall shear stress, pre-existing conditions such as Marfan syndrome or Ehlar-Danlos syndrome and other complications discussed above.

A recent review article from Nienaber et al. (2014) provided guidelines on treatment options: if there are any complications, intervention with TEVAR should be followed where possible, otherwise open surgery is recommended. If there are no complications, medical treatment is the best option. Thus, although expert consensus can provide certain guidelines (Svensson et al. 2008; Erbel et al. 2001; Nienaber et al. 2014; Tsai et al. 2009; Hagan et al. 2000), no clear advice on exactly when and how to intervene is available, or indeed viable. This is partly due to the fact that the mechanical and haemodynamic characteristics of AD are not fully understood (Rajagopal et al. 2007). It may also be the case that due to the vast array of comorbities associated with dissection, the use of statistical analysis is limited as each case is different, and thus patient-specific approaches are necessary in order to improve prognosis for AD (Midulla et al. 2012; Taylor et al. 1999).

1.1.2.5 Biomechanics of Aortic Dissection

In the healthy state, an artery wall has three layers, each with its own functionality. However, these layers behave differently after formation of an AD, due to the excessive damage in their structure. Studies have shown that the outer layer of the FL, which consists of only the adventitia, has approximately a quarter of the vessel thickness compared to a healthy aorta (Shiran et al. 2014). The inner layer of the aortic wall, which forms the intimal flap (IF), is partly made up of intima and media layers. Most of the wall of the TL has similar structure to a healthy artery, except the region that forms the intimal flap. As a result, one could expect that the true lumen would be more stable than the false lumen. However, the pressurised FL frequently compresses the TL, especially during diastole (Yang et al. 2014). Compression of the TL by the FL can lead to downstream organ ischaemia as the upstream vessel is obstructed or collapsed (Criado, 2011), highlighting the complexity of the interactions between the two lumina. Furthermore, the elasticity and distensibility of the dissected vessel wall decreases over time due to the process of fibrosis which leads to an increase in stiffness of the vessel wall (Criado, 2011; Adams et al. 1995; Pyeritz 2000).

1.2 Numerical Modelling of the Cardiovascular System

1.2.1 Background

Computational approaches are becoming a powerful tool in medical research, medical device design, and surgical planning and intervention. Despite continuous improvements in imaging techniques and non-invasive flow measurements, it remains very challenging to acquire in vivo data in patients with sufficient accuracy to fully understand a patient's condition. Modelling approaches such as computational fluid dynamics (CFD), with certain patient-specific inputs (such as geometry from imaging data, heart rate, flow measurements from ultrasound etc.), can yield high spatial and temporal resolution data on parameters such as flow, velocity gradients, pressure, wall shear stress and vessel wall motion, which can assist clinicians in the decision making process (Taylor and Draney 2004). Furthermore, once a model has been developed, it can be modified parametrically, either in such a way as to investigate the haemodynamic effects of various treatment options (Karmonik et al. 2011b; Chen et al. 2013b), or to further investigate the effects of certain parameters and thus to test hypotheses on the formation and progression of the disease (Tsai et al. 2008; Rudenick et al. 2010; Taylor and Draney 2004). Several books (Formaggia et al. 2009; Coveney et al. 2014) have been written on the subject of numerical/computational modelling of the vasculature. Here, a review of the factors most significant to the simulation of type-B AD in a clinical setting will be given.

High spatial resolution imaging techniques such as computed tomography (CT) and three-dimensional magnetic resonance imaging (3D-MRI) provide information on the morphology of the dissected aorta and time-resolved imaging (e.g. using electrocardiograms as a trigger) can yield data on the movement of the wall. Four dimensional MRI (4D-MRI) enables simultaneous quantification of wall motion and measurement of the velocity field, but the technique is prohibitively expensive and its use is not yet widespread. Furthermore, none of the aforementioned methods are able to provide data on the pressure along the dissected region and the false lumen (FL) or have sufficient spatial resolution to accurately calculate the wall shear stress (WSS), which is reported to be the most influential haemodynamic parameter in the growth and rupture of AD (Francois et al. 2013). Furthermore, the limited resolution of velocity measurements using, e.g. 4D-MRI, is particularly problematic when measuring slow flows in the FL, as signal-to-noise ratios are too high for reliable interpretation of the data (Clough et al. 2012).

As with any modelling approach, there must be a trade-off between computational accuracy and efficiency. Furthermore, it should be noted that additional complexity does not necessarily correlate with increased accuracy, particularly when modelling a system in which data for appropriate validation is scarce. In the context of hypothesis testing with regards to developing fundamental understanding of the biomechanics of aortic diseases, the time to run a simulation is not critical. However, for utility as a clinical interventional planning tool (particularly in the context of AD wherein mortality rates increase by the hour), it is also necessary to ensure that a simulation can be completed within a clinically relevant time frame.

Modelling of AD, as with modelling of any biological system, may have one of two aims: either to elucidate aspects of the disease via parametric investigation, or to analyse and plan treatment options of individual patients. The following section summarises reports of both of these functions, and is preceded by a discussion of the modelling assumptions made and their implications for the analysis and applicability of the results.

1.2.2 Approaches to Modelling Aortic Dissection

1.2.2.1 Geometrical Representation of the Anatomy

One of the most critical aspects of AD is the geometry (morphology). Combinations of entry and re-entry tears create unique morphologies for each patient. Parameters such as tear size, orientation, location and number, play an important role on the flow behaviour in AD. Additionally, the inclusion or exclusion of the numerous arteries that branch off the aorta (carotids, sublcavian, iliac, renal, mesenteric etc.) as well as the effects of the rest of the downstream vasculature, are important. Two fundamental approaches have been adopted in the modelling of AD: use of patient-specific geometries and use of idealised geometries.

The use of idealised geometries is most appropriate for parametric analyses on the tear parameters discussed above. Typically, literature values for critical dimensions (such as diameter, arch curvature, tear geometry) are used to create a simplified model (Rudenick et al. 2010). Geometric considerations such as tapering of the aorta are commonly overlooked (Rudenick et al. 2010; Hou et al. 2010; Fan et al. 2010), and some studies even opt to omit the supraaortic branches (those branching off from the aortic arch) (Fan et al. 2010; Tang et al. 2012; Rudenick et al. 2010). However, such idealised models are rare in the literature, perhaps due to the scepticism of the clinical community. Thus, the majority of AD models in the literature are 'patient-specific': which in this context, means derived from clinical imaging data of a given patient.

Reconstruction of 3D geometries from two-dimensional (2D) images is a non-trivial, but commonly practised technique, and will be described in detail in Chap. 2. Briefly, thresholds are applied to a stack of images to separate the vessel section of interest from the background, and a 3D geometry is reconstructed from the planes. The geometry is typically smoothened to remove pixelation effects, and the domain boundaries are trimmed.

In a number of studies, the trimming step includes removal of the supraaortic branches (Cheng et al. 2010, 2013, 2014; Qiao et al. 2014). This was justified as being done in the interest of simplicity (Cheng et al. 2010). However, a critical region in type-B AD is the distal arch, around where the FL is often formed and the entry tear is located. Removing the supraaortic branches significantly alters the local haemodynamics, and so limits the utility of a 'patient-specific' approach.

Furthermore, a significant amount of the flow entering the aortic arch (≈15% in healthy aortae) leaves via these branches. Thus, for a given inlet flow, removal of the branches will artificially increase the flow rate through the descending aorta, particularly in type-B AD, where the increased hydrodynamic resistance of the dissected region can reduce flow to the descending aorta and thus increase the proportional flow through the supraaortic branches. Cheng et al. (2014) applied a modification in order to correct for this by 'scaling' the inlet flow based on the relative magnitude of the flow through the branches compared to the inlet flow at peak systole based on 4D-MRI data. However, this approach tacitly assumes that the phases of the flow in each of the branches is equal to that of the aorta, which is not the case, and thus even with this modification the flow rate through the descending aorta will not be correct.

At the distal end of the descending aorta, the aorta bifurcates into the iliac arteries. Many studies trim the geometry prior to this bifurcation (Cheng et al. 2010; Chen et al. 2013b), whilst others include both the iliac arteries (Tse et al. 2011; Karmonik et al. 2008, 2011b).

In addition to the supraaortic branches, which branch off the aortic arch, there are a number of key arteries that branch off from the descending aorta, including the mesenteric arteries, celiac artery and renal arteries; collectively known as the visceral arteries. These vessels are considerably smaller than the supraaortic branches and are often hard to accurately resolve from imaging data, particularly in the case of AD, wherein the tortuous geometry makes it difficult to distinguish the smaller arteries from the surrounding tissue. As a result, these branches are omitted in the majority of CFD studies of AD, although a few retain some of the branches; Tse et al. (2011) included the renal arteries; Chen et al. (2013) included the celiac artery and Karmonik et al. (2011b) included the renal arteries, celiac artery and superior mesenteric artery. Omission of the visceral arteries is also common in simulations of other aortic diseases, such as aortic coarctation (Kim et al. 2009; Brown et al. 2012), and hyperplasia (Coogan et al. 2011). For abdominal aortic aneurysms, a number of studies omit the branches (Biasetti et al. 2011), while others include all branches in the abdominal aorta only (Frauenfelder et al. 2006; Suh et al. 2010). Sughimoto et al. (2013) modelled the whole aorta, including the arch and all branching arteries, in an investigation of aneurysm repair. Given that the majority of numerical studies of the aorta omit these branches, doing so is generally considered acceptable. Nonetheless, particularly in the context of AD, wherein blockage of these vessels causes malperfusion, it would be advantageous to include the arteries branching from the abdominal aorta, where the quality of the imaging data is sufficiently accurate to enable reconstruction.

The influence of the accuracy of image reconstruction on numerical results has not, to the author's knowledge, been reported in detail. Berthier et al. (2002) showed that reduced order models of coronary arteries drastically altered the results as compared to a realistic reconstruction. Relatively small virtual changes to vessel geometries have also been shown to have a considerable effect on the predicted flow characteristics (Abraham et al. 2008). Indeed, the hydrodynamic resistance of straight tubes varies with the fourth power of the diameter, and thus small changes in geometry might be expected to influence the results. However, it is not feasible in most circumstances to analyse the accuracy of geometric reconstructions quantitatively, and thus it is tacitly assumed that the 3D domain is a sufficiently accurate representation of the geometry of the patient's aorta.

In summary, from a purely morphological point of view, fewer simplifications will lead to a better representation of the haemodynamics in a given geometry. However, for each additional branch included in the model, BCs need to be applied.

1.2.2.2 Boundary Conditions

The appropriate application of BCs in computational models of AD is an essential and often overlooked component of producing realistic results (Grinberg et al. 2009). Accurate data from specific patients at all of the domain boundaries is often unavailable, and thus a number of assumptions must be made. In the case of AD, numerical models are typically 3D representations of the aorta, although this is only a small part of the circulatory system: a complex interactive system, modulated by mechanical, biochemical and neurological signals. Whilst it is impractical to attempt to model the entire vasculature and its regulatory processes, realistic representation of the mechanical (haemodynamic) environment at the interface of the 3D domain is imperative.

In the majority of AD simulations, a prescribed flow rate is used as an inlet boundary condition. Where available, patient-specific inflow BCs are used (Karmonik et al. 2008; Cheng et al. 2014). A recent study reported that directly applying 3D velocities extracted from phase-contrast magnetic resonance imaging (pcMRI) at the inlet for simulations of blood flow in the aorta is the most appropriate inlet boundary condition (Morbiducci et al. 2013), and enables the best prediction of the helical flow structures. However, such data is not available for the majority of AD studies.

A common alternative in the absence of patient specific inflow data is the application of flow rates from previous studies (Chen et al. 2013a, b; Cheng et al. 2010; Tse et al. 2011). While this reduces the 'patient-specific' nature of the simulations, it is a necessary simplification. In order to apply a flow rate at the boundary, the prescribed velocity at each inlet node must be defined. This is generally done by applying a uniform velocity across the inlet (Chen et al. 2013a, b; Cheng et al. 2010, 2013, 2014; Tse et al. 2011; Cheng et al. 2015). Alternatives include mapping parabolic (Poiseuille) or Womersley velocity profiles onto the patient specific geometry. However, it is not clear whether this additional complexity in applying scaled analytical velocity profiles improves accuracy in the aorta (Morbiducci et al. 2013). A similar conclusion was drawn in an intracranial aneurysm (Marzo et al. 2009) and a carotid bifurcation (Campbell et al. 2012).

Applying the BCs at the system outlets (branched arteries and the distal abdominal) is a more complex task, as the flow rate through each outlet will be dependent on the inlet and all of the other outlets. The most common approaches are application of a constant pressure (usually zero) or 'flow splitting', based on literature values of flow proportions through specific branches.

The application of zero pressure at the outlets is the simplest BC (Cheng et al. 2010, 2013, 2014; Chen et al. 2013a; Karmonik et al. 2011b; Qiao et al. 2014). In rigid wall simulations, the value of zero (as compared to a different constant pressure) is not important as it is the pressure drop that drives the flow, rather than the absolute pressure. Nevertheless, throughout the cardiac cycle, the pressures at all outlet branches vary in a way that is comparable to the inlet, but are phase shifted and modified in terms of magnitude. Therefore at a given time in the cardiac cycle, the pressure gradient across the geometry may be only a few mmHg (Olufsen et al. 2000). By applying a constant pressure outlet, if a flow inlet is used, the pressure at the inlet will be very low, whereas if a pressure inlet is used, the flow through the system will be excessively high. It is thus clearly necessary to use more sophisticated BCs, in order to produce results that are representative of physiological haemodynamics. An alternative option is to prescribe pressure waves from other studies. In particular, the one-dimensional modelling study of Olufsen et al. (2000) provided pressure waves at various locations, supported by comparison with the flow from MRI data. A number of studies have directly applied these generic pressure waves to the outlet in simulations of AD (Chen et al. 2013b; Tse et al. 2011; Fan et al. 2010). However, it should be noted that healthy pressure waves in the thoracic aorta may differ significantly from those in AD.

An alternative approach commonly used for the supraaortic branches is to predefine a proportion of the flow through each branch, in combination with a prescribed pressure wave (Chen et al. 2013b; Tse et al. 2011), or zero pressure (Chen et al. 2013a; Cheng et al. 2015), at the remaining outlet(s). The proportions are often reported to be 5% for each of the branches, and are credited to Shahcheraghi et al. (2002) who in turn got the data from a textbook from 1972 (Middleman 1972). This approach may appear preferential to constant pressure BCs, but the flow rates must be calculated *a priori*. As such, they are calculated by taking 5% of the inlet flow waves, and are thus exactly in phase with the inlet flow, which is not physiologically accurate. Cheng et al. (2015) instead split 30% of the flow between the supraaortic branches based on cross-sectional area, but the phase issue was not resolved.

In order to address these issues, a considerable amount of work has been done on 'zero-dimensional modelling' and its coupling to the 3D domain outlet boundaries to provide a dynamic representation of the influence of the downstream vasculature (Kim et al. 2009; Brown et al. 2012; Formaggia et al. 2009). This approach uses hydraulic-electrical analogue circuits to model resistive and compliant aspects of the vasculature, and is reviewed in detail by Shi et al. (2011). More details will be given on this approach in Chap. 2.

To the best of the author's knowledge, in every simulation of AD to date (with the exception of Qiao et al. (2014)), a rigid wall with a no-slip condition has been used. The validity of this approach is supported by reports of reduced vessel distensibility in AD (Chen et al. 2013b; Cheng et al. 2010), citing the clinical imaging study of Ganten et al. (2009). However, in that study the reported reduction was just 12%, when compared to 'healthy subjects', which in the absence of any statistical analysis, may be within the measurement resolution/variability within the sample populations. Other studies justify the rigid wall modelling approach by reporting that little wall movement was observed in time resolved imaging data specific to the study (Chen et al. 2013a; Karmonik et al. 2011b, 2012a). It has also been pointed out that the complexities involved in modelling the aortic wall will be exacerbated in the context of AD, wherein the FL and IF wall properties are significantly different to those of a healthy aorta, and change over time (Cheng et al. 2010). The presence of the intimal flap in AD additionally makes numerical modelling and discretisation of the domain more complex relative to healthy aortae. Only a single paper has reported on the use of fluid-structure interaction (FSI) methodologies to model wall motion in a simplified model of AD (Qiao et al. 2014). However, by using a linear elastic model for the vessel wall with a Young's modulus of 100 MPa (as compared to normal values in the range 0.4–6 MPa (Crosetto et al. 2011; Khanafer and Berguer 2009; Gao et al. 2006)) the displacements observed were minimal. Although the descending aorta moved by more than 1 mm in certain circumstances, this was essentially rigid body motion, and the relative displacement of the intimal flap had a maximum of $\approx 200\,\mu$m. Thus improved FSI studies that are able to reproduce the motion observed in imaging studies is still required (Ganten et al. 2009; Karmonik et al. 2012a). It remains to be seen how important it is to include wall motion when modelling AD, and this issue will be discussed and investigated further in Chap. 5.

1.2.2.3 Fluid Modelling

Blood is a complex fluid made up of red blood cells, white blood cells and platelets suspended in plasma. The latter is approximately Newtonian, but the suspended elements alter the fluid properties through deformation, aggregation and irreversible coagulation (thrombosis) (Popel and Johnson 2005). Furthermore, the viscosity is dependent on the local concentration of RBCs, which varies throughout the vasculature and within individual vessels (Jung and Hassanein 2008; Popel and Johnson 2005).

The majority of simulations of AD make the assumption that the fluid is Newtonian, whereas blood (when measured in constant shear rheometers) exhibits non-Newtonian properties. Blood is shear-thinning, meaning that as the shear rates increase, the viscosity of the fluid decreases (Levick 2009). This occurs as a results of disaggregation and deformation of RBCs. However, several classic textbooks report that it is sufficient to assume that the fluid in large vessels is Newtonian (Fung 1997; Pedley 1980). A number of studies have proposed and investigated various non-Newtonian models for blood flow in large arteries such as the carotid (Gijsen et al. 1999; Razavi et al. 2011) or coronary arteries (Soulis et al. 2008; Johnston et al. 2004). Jung and Hassanein (2008) used a 3-phase blood viscosity model that accounts for spatial distributions of RBCs as well as shear rate. However, there remains little consensus as to the most appropriate model for blood, perhaps due to the absence of sufficiently accurate experimental data for validation of numerical models. Nonetheless, simple non-Newtonian models, such as the Carreau-Yasuda model

(Gijsen et al. 1999) are expected to improve the efforts at modelling blood. Cheng et al. (2010) used the Quemada viscosity model (Quemada 1977, 1978a, b), which is both haematocrit and shear dependent, although they do not report what parameters were used, and thus it must be assumed that the haematocrit component was considered to be constant. Hou et al. (2010) used a three-phase model similar to Jung and Hassanein (2008) in an idealised model of an AD. The approach uses a Carreau-Yasuda model with empirical shear and haematocrit-dependent viscosity and transport equations for the RBCs to discern local haematocrit. They observed separation of RBCs in the false lumen, with an increase in haematocrit in the distal FL and a decrease in the proximal FL. Despite the potential for this method to provide further insight into the role of haemorheology on haemodynamics in AD, this work has not been followed up, perhaps due to the absence of suitable experimental data for validation purposes.

Another key issue in the modelling of blood flow in the aorta is whether to model the fluid as laminar, transitional or turbulent. Whereas laminar flow is deterministic, when flow becomes turbulent, random fluctuations in velocity occur and turbulent eddies form and dissipate energy in the flow. Turbulent flow is thus less 'efficient' and the vasculature appears to have developed in such a way that healthy aortic flow is just subcritical (Fung 1997) (i.e. exhibiting as high flow rates as possible without inducing turbulence). The state of turbulence is often described in terms of the Reynold's number Re, which is the ratio of inertial to viscous forces. In steady pipe flow, the transition to turbulence occurs at a critical Reynold's number of around 2000. However, as turbulence takes time to develop, pulsatile flow has a much higher critical Reynold's number, Re_c, which increases with the Womersley number (the ratio of transient inertial forces to viscous forces) (Fung 1997). Reynold's decomposition is a method for the analysis of turbulent flows, in which the velocity vector

$$\mathbf{U} = \begin{pmatrix} u \\ v \\ w \end{pmatrix} \tag{1.1}$$

is separated into mean ($\overline{\mathbf{U}}$), periodic ($\tilde{\mathbf{U}}$) and random fluctuation ($\mathbf{U}'$) components (Munson et al. 1994; Poelma et al. 2015):

$$\mathbf{U}(x, y, z, t) = \overline{\mathbf{U}(x, y, z)} + \tilde{\mathbf{U}}(x, y, z, t) + \mathbf{U}'(x, y, z, t) \tag{1.2}$$

The turbulent kinetic energy (at a given location and time), k, i.e. the energy associated with turbulent eddies, can be calculated according to the mean of the turbulent normal stresses (Versteeg and Malalasekera 2007)

$$k = \frac{1}{2}\left(\overline{u'^2} + \overline{v'^2} + \overline{w'^2}\right) \tag{1.3}$$

1.2 Numerical Modelling of the Cardiovascular System

The turbulence intensity is a commonly used measure of the amount of turbulence, and can be calculated according to (Cheng et al. 2010, 2014):

$$Tu(x, y, z, t) = \frac{\sqrt{\frac{2}{3}k}}{|\mathbf{U}(x, y, z, t)|} \quad (1.4)$$

In regions where the velocity is very small, a large Tu may be calculated, even though the energy associated with turbulent flow and thus the influence of the turbulence, may be small. Alternatively, the numerator of Eq. 1.4 provides an indication of the extent of turbulence.

A number of CFD studies report Re_c values in the range an 2700–15000 (Fung et al. 2008; Tse et al. 2011), along with maximum Reynolds numbers up to 3700, and thereby justify use of a laminar model. Other studies simply refer to classic texts such as Fung (1997), stating that flow in large arteries is laminar (Cheng et al. 2008; Chen et al. 2013b; Fan et al. 2010; Qiao et al. 2014). However, although blood flow is generally laminar, turbulent and transitional flow has been observed in the healthy ascending aorta, and is more prevalent in the presence of aortic disease (Tan et al. 2009; Khanafer and Berguer 2009).

Hence, a number of AD studies have utilised turbulence modelling. Tang et al. (2012) used the $k - \varepsilon$ model, which is perhaps the most commonly used turbulence model (Versteeg and Malalasekera 2007). However, the $k - \varepsilon$ model requires special treatment to accurately resolve the flow near the walls. Wilcox's $k - \omega$ model uses turbulence frequency, rather than dissipation rate as a length scale (Versteeg and Malalasekera 2007). In doing so the near-wall modelling capabilities are improved, but accuracy is sacrificed in the free-stream, wherein division by zero in the equations must be avoided by application of an arbitrary small value for ω, which in turn affects all of the results (Menter 1992). Menter (1992) proposed a hybrid 'Shear-Stress Transport' (SST) model, which utilises the $k - \omega$ model near the wall and the $k - \varepsilon$ model away from the wall. Chen et al. (2013b) used the SST model for turbulence in their simulations of AD, and found that although the laminar model predicted a similar distribution of wall shear stresses, the magnitudes were under-predicted when turbulence was not accounted for.

Tan et al. (2008) compared SST to a variant of the SST model, SST Tran, which is designed to better predict turbulence in transitional flows (Menter et al. 2006). On comparison with experimental data in an idealised stenosis, they concluded that the SST Tran model reproduced experimental data better than the SST model. The latter over predicted turbulence intensity, although on inspection neither model achieved a particularly good reproduction of Tu (Tan et al. 2008). A number of AD studies have thus adopted the SST Tran turbulence model in the modelling of AD. Cheng et al. (2010), used the SST Tran model in a simplified AD geometry (with the aortic branches removed) and observed a considerable amount of turbulence around the first tear in both the true and false lumina. However, it should be noted that their geometry had a particularly severe coarctation and, by removing the branches, they artificially increased the proportion of the inlet flow passing through the coarcted region, although the flow rate was scaled down to try to correct for this. They observed a maximum turbulence intensity of 70% in the systolic decelerating phase, due to the jet produced by the severe coarctation. However, use of a non-Newtonian model reduced the turbulence in the false lumen. In a single patient-longitudinal study, Cheng et al. (2013) observed that a site of high Tu was collocated with the site of thrombosis in a follow up. In a subsequent study, Cheng et al. (2014) reported much lower levels of turbulence, which were predominantly located in the false lumen.

Tan et al. (2009) compared laminar and SST Tran models to MRI data in a thoracic aneurysm and reported that the SST Tran model was superior in terms of reproducing velocity profiles, particularly at peak systole. However, the analysis was qualitative and a degree of subjectivity remains. Cheng et al. (2014) compared the SST model to MRI data in more detail and showed quite good reproduction of all three velocity components at four points in the cardiac cycle, although no statistical evidence was provided.

An alternative to using turbulence modelling as described above is to use a sufficiently fine computational mesh and time step that all of the turbulent length and times scales can be resolved, resulting in a so called direct numerical simulation (DNS) (Moin and Mahesh 1998). For example, in a model of an intracranial aneurysm (using rigid wall and Newtonian fluid), Valen-Sendstad et al. (2011) used 3 million mesh elements and a timestep of 0.025 ms and observed transitional flow, in agreement with experimental observations. A subsequent paper from the same authors showed, even with a steady input flow, transitional flow occurred in 5 of 12 intracranial aneurysms considered, which also turned out to be the aneurysms that ruptured, indicating a role for turbulence in such situations (Valen-Sendstad et al. 2013). Another study used around 24 million mesh elements in models of the carotid siphon, and reported transitional flow to varying degrees in all 5 geometries (Valen-Sendstad et al. 2014). Poelma et al. (2015) used a similar modelling approach with 5.65 million elements in an abdominal aortic aneurysm. They found that a 5 ms timestep was sufficient to resolve the transitional flow in the geometry. The key conclusion from this work was that, in sufficiently large geometries, wherein transients do not decay between cycles, the partially decayed turbulence acts as a random set of initial conditions such that the flow patterns can change significantly between subsequent cycles. As a corollary, using a single cycle to analyse results such as WSS indices may yield results that are not representative of the average WSS in the geometry, and thus multiple cycles should be analysed. These super-high resolution modelling approaches are capable of modelling turbulence better than the turbulence models described above, but have the disadvantage that the computational cost is extremely high.

In summary, there is, at present, no sufficiently rigorous analysis to validate the most appropriate choice of turbulence or viscosity models in the context of AD.

Finally, in addition to numerical modelling of AD, a number of studies have used in vitro 'phantoms' (models) of idealised ADs with the aim of 'validating' computational methods (Rudenick et al. 2010, 2013; Tsai et al. 2008; Soudah et al. 2015). Rudenick et al. (2010) created a simplified, rigid-walled model of a dissected aorta (without branches, but with the aortic arch) and used water as a fluid along with clinical imaging modalities (ultrasound and Doppler) to analyse the flow.

They additionally carried out CFD simulations on the same geometry and compared their results with in vivo data, and reported good agreement. In a subsequent in vitro study, they further simplified the geometry (thereby removing the helical flow introduced by the arch) but looked at different tear configurations (proximal only, distal only, proximal and distal) at two different sizes, and used compliant walls. They observed that all tears acted as both entry and re-entry, with fluid entering the FL in systole and exiting in diastole through all tears. Soudah et al. (2015) simulated the same experimental data (Rudenick et al. 2013) and 'validated' their computational model in cases where wall compliance had little effect. However, when either of the tears was occluded the predictions of the pressure in the FL were very poor. Tsai et al. (2008) also investigated tear configuration with an in vitro geometry.

While the above studies have attempted to use in vitro approaches to improve knowledge, the use of clinical measurement tools (ultrasound and Doppler flowmetry, catheter pressure measurements) rather than engineering techniques such as laser Doppler velocimetry or particle image velocimetry, means that the experimental measurements are not significantly better than those that can be obtained in vivo. The main advantage of in vitro approaches is that they can more accurately measure blood flow, and thus validate the choice of turbulent and viscosity models. To the best of the author's knowledge, this has not been done, as water was used instead in the aforementioned studies. The current in vitro AD models are thus still developmental and have only compared numerical simulations of water to experimental measurements of water in a model geometry.

1.2.3 Objectives of Aortic Dissection Modelling

Numerical modelling of AD has been carried out with a number of different stated objectives. Karmonik et al. (2008) published one of the first CFD studies of AD, as a proof of concept. A number of other studies have followed, using simulations as basic 'haemodynamic investigations', with considerations of turbulence and viscosity (Cheng et al. 2010), TL-FL flow distribution (Chen et al. 2013a), or comparison of in vitro, in vivo and *in silico* (Rudenick et al. 2010). Other studies have more specific aims, such as validating simulations by comparison with MRI (Cheng et al. 2014) or investigating the role of blood characteristics (Hou et al. 2010).

1.2.3.1 Tear Size and Location

A number of studies have investigated the role of tear-size and location on haemodynamics. Fan et al. (2010) used an idealised geometry to study how tear size affected the flow in the FL. It was hypothesised that regions of negligible flow will eventually form thrombosis and thus the location of the 'cut-off plane' for flow (presumably where the velocity was lower than a certain threshold) in the FL was used as indicative of the amount of thrombosis. Increasing tear size and moving the re-entry tear closer to the entry tear decreased the amount of FL flow, as did increasing the diameter of the FL, thereby increasing the extent of thrombosis according to their hypothesis. However, no attempt was made to model thrombosis or compare the results to clinical data. Tang et al. (2012) used a smoothened and simplified model geometry to analyse a number of factors, including the size of the aneurysm, the blood pressure and location of tears. The total force acting on the FL was used as an index, under the hypothesis that this would be indicative of further enlargement. However, given that the normal force is the integral of the pressure over the area of the FL, and this greatly exceeds the shear force, the results that increasing pressure and increasing the area resulted in a greater force are not very insightful. Cheng et al. (2013) analysed four patients and compared tear height (length) and width to the proportion of flow in the FL. They found that larger tears correspond to a greater proportion of FL flow, as did entry tears located closer to the arch top, although a larger sample size would be required to obtain statistical significance for these trends (Cheng et al. 2013). In a subsequent study, Cheng et al. (2015) observed the same trend in tear size with four surgically, and four medically treated patients.

1.2.3.2 Tear Configuration

The tear configuration has been studied by a number of researchers. Tear configuration is generally analysed in an AD geometry with an entry and re-entry tear. With both tears open, the system of a patent false lumen is modelled. Blocking the re-entry tear represents partial thrombosis. Occlusion of the entry tear can be used to simulate TEVAR treatment, and occluding both tears can be used to model complete thrombosis, or a more extensive TEVAR operation. Tsai et al. (2008) used an idealised in vitro model of an AD, and modelled the first three scenarios described above. They analysed the difference in pressure between the FL (P_{FL}) and TL (P_{TL}) and found a small but significant increase from $P_{FL}/P_{TL} \approx 1$ to $P_{FL}/P_{TL} \approx 1.05$ after TEVAR treatment and $P_{FL}/P_{TL} \approx 1.08$ for a partially thrombosed FL. Rudenick et al. (2013) also considered the same three scenarios, as well as varying the tear size, and found that large tears reduced the FL pressure ratio from $P_{FL}/P_{TL} \approx 1.0$ to 0.93. The compliance of their phantom was also such that both tears acted as entry and exit tears. Chen et al. (2013b) considered alternative tear configurations in a geometry with multiple tears and concluded that effective treatment would require covering all of the tears. Karmonik et al. (2011b) considered all four scenarios, i.e. occluding either tear or both tears. They also found the largest intraluminal pressure gradient in the 'partial thrombosis' case, although the pressure was similar to the patent lumen case. Complete removal of the IF, as a model of surgical fenestration, reduced the pressure in the combined lumen to a third of what was observed in the patent lumen case. Another study considering treatment options was that of Qiao et al. (2014), who modelled the effect of bypassing the dissected region entirely either from the ascending aorta, or from where the left subclavian artery would have been, if they had not removed it from the model. However, as a result of the removal of the branches, it is not possible to know how these extreme treatment options might affect the flow through the supraaortic branches, which may have been a useful result from the study.

1.2.3.3 Follow up and Multiple Patient Studies

Due to the relative rarity of AD, a large number of AD studies consider a single patient, at a single time-point only (Cheng et al. 2010; Chen et al. 2013a; Karmonik et al. 2011b). However, a number of studies have been able to get follow up scans on the same patient. Karmonik et al. (2012) presented data based on follow up scans on a patient who exhibited aneurysmal dilatation by 26%. The corresponding pressure and WSS in the FL were reduced by over 50%. In another study, Karmonik et al. (2011a) reported on follow up of a patient who underwent endovascular aortic repair surgery and found that the treatment had reduced the wall shear stresses and the pressure acting on the vessel wall. Tse et al. (2011) analysed scans from a patient suffering a type-B AD before and after the formation of a dissecting aneurysm, but were only able to identify that the site of aneurysmal formation was collocated with a region of intraluminal pressure drop of 1.5 mmHg, which is not particularly high. Cheng et al. (2013) reported on 8 scans over the period of a year, and observed that the site of thrombosis in the FL was collocated with regions of high turbulence intensity and relative residence time (RRT: a parameter that is used to represent the amount of time that blood particles spend in a localised region). Chen et al. (2013b) considered a patient who was treated medically on initial examination and four years later. In the absence of patient-specific BCs, they used the same flow rate but scaled the inlet flow so as to achieve a 'healthy' pressure drop across the system. Other than the effect that reduced inlet flow rate had on the magnitude of the velocities (which were also reduced), very little difference was observed between the two cases. Cheng et al. (2013) published one of the few papers to report on multiple patients (four), and a subsequent study with 8 patients, comparing medically and surgically managed cases (Cheng et al. 2015). Although even more subjects would be required to establish statistical significance, the use of multiple patients is clearly advantageous in allowing more general trends to be observed.

In the long term, it will be necessary to compile longitudinal data sets on a large number of patients, and combine them with 'blind' simulations, wherein researchers are given data before a disease development or treatment, and tasked with predicting the outcome. This will allow the community to ascertain what the most clinically relevant parameters output by CFD simulations are, and could provide guidelines for what assumptions are appropriate.

1.3 Objectives of the Present Research

The research carried out in this thesis aims to provide additional insight into the modelling of AD. Based on the preceding review of the literature, a number of key points that need to be addressed have been identified. Three objectives specifically will be investigated in this thesis:

- **Dynamic boundary conditions**: To the best of the author's knowledge, previous CFD models of AD have ubiquitously predefined the BCs. However, interactions between the downstream vasculature and the 3D domain are expected to influence the haemodynamics. The first objective of this research will be to develop and implement a methodology for coupling dynamic BCs to the 3D domain of a patient-specific dissected aorta.
- **Virtual stenting**: Although a number of studies have investigated virtual stenting, the influence of dynamic BCs on the predictions made by such simulations is not known. Additionally, improved techniques for analysing the efficacy of the treatments would render such approaches more valuable. The second objective of this research is to analyse the influence of virtual treatments with dynamic BCs, and to develop new approaches to extract useful conclusions from the results.
- **Wall motion**: There are as yet no FSI studies reported in the literature that are capable of producing the wall motions observed in AD. The impact that this has on computational results has been speculated in many papers, but FSI simulations are required in order to investigate the influence of wall motion on CFD predictions in AD. The third objective of this research is to provide the first numerical results on the influence of wall motion on the clinical value of CFD simulations of AD.

All of the above objectives will be investigated with a view to developing tools that may be transferred to the clinic, and thus the computational efficiency as well as accuracy of the simulations are considered.

1.4 Outline of the Thesis

Chapter 2 will provide a background to patient-specific modelling. An introduction to computational fluid dynamics will be given. This will be followed by a description of clinical imaging methodologies and the extraction of the 3D domain and meshing for use in numerical studies. The mathematical background and relevant literature on coupling 3D-0D models for AD will be given, followed by supporting evidence for the use of such approaches. The chapter will conclude with a brief description of relevant aspects of finite element modelling.

A simulation of a patient-specific dissected aorta will be investigated in Chap. 3. The dynamic BCs will be used at the domain's outlets and a novel technique for tuning the parameters to invasive patient-specific pressure data will be presented, implemented and analysed. A detailed description of the haemodynamics in the dissected aorta will be followed by a thorough analysis of the sensitivity of the results to model parameters, turbulence and mesh refinement.

In Chap. 4, two virtual stenting operations will be described, representing two different treatments for the patient. The dynamic BCs derived in the preceding chapter will be used to provide comparable results between cases. In addition to pressure, flow and WSS patterns, energy loss and statistical distributions of WSS indices will be given.

Chapter 5 will present a model that takes into account the distensibility of the vessel wall, dynamic BCs, non-Newtonian viscosity and turbulence. The data will be compared to a rigid wall simulation and analysed in the context of whether the additional computational time required for these simulations is justified.

The thesis will conclude with Chap. 6, which will summarise the main results from the research and discuss their implications for future studies.

References

Abraham, J. P., Sparrow, E. M., & Lovik, R. D. (2008). Unsteady, three-dimensional fluid mechanic analysis of blood flow in plaque-narrowed and plaque-freed arteries. *International Journal of Heat and Mass Transfer, 51*(23–24), 5633–5641.

Adams, J. N., Brooks, M., Redpath, T. W., Smith, F. W., Dean, J., Gray, J., et al. (1995). Aortic distensibility and stiffness index measured by magnetic resonance imaging in patients with Marfan's syndrome. *British Heart Journal, 73*(3), 265–269.

Alfonso, F., Almeria, C., Fernandez-Ortiz, A., Segovia, J., Ferreiros, J., Goicolea, J., et al. (1997). Aortic dissection occurring during coronary angioplasty: Angiographic and transesophageal echocardiographic findings. *Catheterization and Cardiovascular Diagnosis, 42*(4), 412–415.

Aziz, F., Penupolu, S., Alok, A., Doddi, S., & Abed, M. (2011). Peripartum acute aortic dissection: A case report & review of literature. *Journal of Thoracic Disease, 3*(1), 65–67.

Bastien, M., Dagenaisa, F., Dumonta, E., Vadeboncoeura, N., Diona, B., Royera, M., et al. (2012). Assessment of management of cardiovascular risk factors in patients with tho-racic aortic disease. *Clinical Methods and Pathophysiology, 17*(6), 235–242.

Beighton, P., Paepe, A. D., Steinmann, B., Tsipouras, P., & Wenstrup, R. J. (1998). Ehlers-Danlos syndromes: Revised nosology, Villefranche, 1997. *American Journal of Medical Genetics, 77*(1), 31–37.

Bernard, Y., Zimmermann, H., Chocron, S., Litzler, J. F., Kastler, B., Etievent, J. P., et al. (2001). False lumen patency as a predictor of late outcome in aortic dissection. *The American Journal of Cardiology, 87*(12), 1378–1382.

Berthier, B., Bouzerar, R., & Legallais, C. (2002). Blood flow patterns in an anatomically realistic coronary vessel: Influence of three different reconstruction methods. *Journal of Biomechanics, 35*(10), 1347–1356.

Biasetti, J., Hussain, F., & Gasser, T. C. (2011). Blood flow and coherent vortices in the normal and aneurysmatic aortas: A fluid dynamical approach to intra-luminal thrombus formation. *Journal of The Royal Society Interface, 8*(63), 1449–1461.

Booher, A. M., Isselbacher, E. M., Nienaber, C. A., Trimarchi, S., Evangelista, A., Montgomery, D. G., et al. (2013). The IRAD classification system for characterizing survival after aortic dissection. *The American Journal of Medicine, 126*(8), 730.e19–24.

Bozinovski, J., & Coselli, J. S. (2008). Outcomes and survival in surgical treatment of descending thoracic aorta with acute dissection. *The Annals of Thoracic Surgery, 85*(3), 965–971.

Braverman, A. C. (2011). Aortic dissection: Prompt diagnosis and emergency treatment are critical. *Cleveland Clinic Journal of Medicine, 78*(10), 685–696.

Brown, A. G., Shi, Y., Marzo, A., Staicu, C., Valverde, I., Beerbaum, P., et al. (2012). Accuracy vs. computational time Translating aortic simulations to the clinic. *Journal of Biomechanics, 45*(3), 516–523.

Burchell, H. B. (1955). Aortic dissection (dissecting hematoma; dissecting aneurysm of the aorta). *Circulation, 12*(6), 1068–1079.

Campbell, I. C., Ries, J., Dhawan, S. S., Quyyumi, A. A., Taylor, W. R., & Oshinski, J. N. (2012). Effect of inlet velocity profiles on patient-specific computational fluid dynamics simulations of the carotid bifurcation. *Journal of Biomechanical Engineering, 134*(5), 051001–0510018.

Capoccia, L., & Riambau, V. (2014). Current evidence for thoracic aorta type B dissection management. *Vascular, 22*(6), 439–447.

Chen, D., ller Eschner, M. M., von Tengg-Kobligk, H., Barber, D., Bockler, D., Hose, R., et al. (2013a). A patient-specific study of type-B aortic dissection: evaluation of true-false lumen blood exchange. *BioMedical Engineering OnLine, 12*, 65.

Chen, D., Müller-Eschner, M., Kotelis, D., Böckler, D., Ventikos, Y., & von Tengg-Kobligk, H. (2013b). A longitudinal study of Type-B aortic dissection and endovascular repair scenarios: Computational analyses. *Medical Engineering and Physics, 35*(9), 1321–1330.

Chen, D., Müller-Eschner, M., Rengier, F., Kotelis, D., Böckler, D., Ventikos, Y., et al. (2013c). A preliminary study of fast virtual Stent-Graft deployment: Application to stanford type B aortic dissection. *International Journal of Advanced Robotic Systems, 10*, 154.

Cheng, S. W. K., Lam, E. S. K., Fung, G. S. K., Ho, P., Ting, A. C. W., & Chow, K. W. (2008). A computational fluid dynamic study of stent graft remodeling after endovascular repair of thoracic aortic dissections. *Journal of Vascular Surgery, 48*(2), 303–310.

Cheng, Z., Tan, F. P. P., Riga, C. V., Bicknell, C. D., Hamady, M. S., Gibbs, R. G. J., et al. (2010). Analysis of flow patterns in a patient-specific aortic dissection model. *Journal of Biomechanical Engineering, 132*(5), 051007.

Cheng, Z., Riga, C., Chan, J., Hamady, M., Wood, N. B., Cheshire, N. J., et al. (2013). Initial findings and potential applicability of computational simulation of the aorta in acute type B dissection. *Journal of Vacscular Surgery, 57*(2), 35S–43S.

Cheng, Z., Juli, C., Wood, N. B., Gibbs, R. G. J., & Xu, X. Y. (2014). Predicting flow in aortic dissection: Comparison of computational model with PC-MRI velocity measurements. *Medical Engineering and Physics, 36*(9), 1176–1184.

Cheng, Z., Wood, N. B., Gibbs, R. G. J., & Xu, X. Y. (2015). Geometric and Flow Features of Type B Aortic Dissection: Initial Findings and Comparison of Medically Treated and Stented Cases. *Annals of biomedical engineering, 43*(1), 177–189.

Clough, R. E., Waltham, M., Giese, D., Taylor, P. R., & Schaeffter, T. (2012). A new imaging method for assessment of aortic dissection using four-dimensional phase contrast magnetic resonance imaging. *Journal of Vacscular Surgery, 55*(4), 914–923.

Coady, M. A., Rizzo, J. A., & Elefteriades, J. A. (1999). Pathologic variants of thoracic aortic dissections. Penetrating atherosclerotic ulcers and intramural hematomas. *Cardiology Clinics, 17*(4), 637–657.

Coogan, J. S., Chan, F. P., Taylor, C. A., & Feinstein, J. A. (2011). Computational fluid dynamic simulations of aortic coarctation comparing the effects of surgical- and stent-based treatments on aortic compliance and ventricular workload. *Catheterization and Cardiovascular Interventions, 77*(5), 680–691.

Coveney, P., Díaz-Zuccarini, V., Hunter, P., & Viceconti, M. (2014). *Computational biomedicine*. Oxford University Press.

Crawford, E. (1990). The diagnosis and management of aortic dissection. *Journal of the American Medical Association, 264*(19), 2537–2541.

Criado, F. J. (2011). Aortic dissection: A 250-year perspective. *Texas Heart Institute Journal,38*(6), 694–700.

Crosetto, P., Reymond, P., Deparis, S., & Kontaxakis, D. (2011). Fluid-structure interaction simulation of aortic blood flow. *Computers and Fluids, 43*, 46–57.

Daniel, J. C., Huynh, T. T., Zhou, W., Kougias, P., El Sayed, H. F., Huh, J., et al. (2007). Acute aortic dissection associated with use of cocaine. *Journal of Vascular Surgery, 46*(3), 427–433.

De Bakey, M. E., Henly, W. S., Cooley, D. A., Morris, G. C. J., Crawford, E. S., & Beall, A. C. J. (1965). Surgical management of dissecting aneurysms of the aorta. *The Journal of Thoracic and Cardiovascular Surgery, 49*, 130–149.

References

De Rango, P., & Estrera, A. (2011). Uncomplicated type B dissection: Are there any indications for early intervention? *The Journal of Cardiovascular Surgery*, *52*(4), 519–528.

De Bakey, M. E., Cooley, D. A. & Creech, O. (1955). Surgical considerations of dissecting aneurysm of the aorta. *Annals of Surgery*, *142*(4), 586–610–discussion–611–612.

Delsart, P., Beregi, J.-P., Devos, P., Haulon, S., Midulla, M., & Mounier-Vehier, C. (2013). Thrombocytopenia: an early marker of late mortality in type B aortic dissection. *Heart and Vessels*, *29*(2), 220–230.

Dietz, H. C., Cutting, G. R., Pyeritz, R. E., Maslen, C. L., Sakai, L. Y., Corson, G. M., et al. (1991). Marfan syndrome caused by a recurrent de novo missense mutation in the fibrillin gene. *Nature*, *352*(6333), 337–339.

Drake, R., Vogl, A., & Mitchell, A. (2010). *Gray's anatomy for students* (2nd ed.). Churchill Livingstone: Elsevier.

Erbel, R., Oelert, H., Meyer, J., Puth, M., Mohr-Katoly, S., Hausmann, D., et al. (1993). Effect of medical and surgical therapy on aortic dissection evaluated by transesophageal echocardiography. Implications for prognosis and therapy. The European Cooperative Study Group on Echocardiography. *Circulation*, *87*(5), 1604–1615.

Erbel, R., Alfonso, F., Boileau, C., Dirsch, O., Eber, B., Haverich, A., et al. (2001). Diagnosis and management of aortic dissection task force on aortic dissection, European Society of Cardiology. *European Heart Journal*, *22*(18), 1642–1681.

Erbel, R., Aboyans, V., Boileau, C., Bossone, E., Bartolomeo, R. D., Eggebrecht, H., et al. (2014). 2014 ESC Guidelines on the diagnosis and treatment of aortic diseases: Document covering acute and chronic aortic diseases of the thoracic and abdominal aorta of the adult * The Task Force for the Diagnosis and Treatment of Aortic Diseases of the European Society of Cardiology (ESC). *European Heart Journal*, *35*(41), 2873–2926.

Estrera, A. L., Miller, C. C., Safi, H. J., Goodrick, J. S., Keyhani, A., Porat, E. E., et al. (2006). Outcomes of medical management of acute type B aortic dissection. *Circulation*, *114*(Suppl 1), I384–I389.

Fan, Y., Wing-Keung, S., Kai-Xiong, Q., & Chow, K.-W. (2010). Endovascular repair of type B aortic dissection: A study by computational fluid dynamics. *Journal of Biomedical Science and Engineering*, *3*, 900–907.

Fattori, R., Tsai, T. T., Myrmel, T., Evangelista, A., Cooper, J. V., Trimarchi, S., et al. (2008b). Complicated acute type B dissection: Is surgery still the best option? *JACC: Cardiovascular Interventions*, *1*(4), 395–402.

Fattori, R., Botta, L., Lovato, L., Biagini, E., Russo, V., Casadei, A., et al. (2008a). Malperfusion syndrome in type B aortic dissection: Role of the endovascular procedures. *Acta Chirurgica Belgica*, *108*(2), 192.

Fattori, R., Cao, P., De Rango, P., Czerny, M., Evangelista, A., Nienaber, C., et al. (2013). Interdisciplinary expert consensus document on management of type B aortic dissection. *Journal of the American College of Cardiology*, *61*(16), 1661–1678.

Fisher, A., & Holroyd, B. R. (1992). Cocaine-associated dissection of the thoracic aorta. *The Journal of Emergency Medicine*, *10*(6), 723–727.

Formaggia, L., Quarteroni, A. M., & Veneziani, A. (2009). *Cardiovascular mathematics: modeling and simulation of the circulatory system*. Springer.

Francois, C. J., Markl, M., Schiebler, M. L., Niespodzany, E., Landgraf, B. R., Schlensak, C., et al. (2013). Four-dimensional, flow-sensitive magnetic resonance imaging of blood flow patterns in thoracic aortic dissections. *The Journal of Thoracic and Cardiovascular Surgery*, *145*(5), 1359–1366.

Frauenfelder, T., Lotfey, M., Boehm, T., & Wildermuth, S. (2006). Computational fluid dynamics: Hemodynamic changes in abdominal aortic aneurysm after Stent-Graft implantation. *CardioVascular and Interventional Radiology*, *29*(4), 613–623.

Fronek, K., & Zweifach, B. (1975). Microvascular pressure distribution in skeletal muscle and the effect of vasodilation. *American Journal of Physiology*, *228*(3), 791–796.

Fung, Y. (1997). *Biomechanics: circulation* (2nd ed.). Springer.

Fung, G. S. K., Lam, S. K., Cheng, S. W. K., & Chow, K. W. (2008). On stent-graft models in thoracic aortic endovascular repair: A computational investigation of the hemodynamic factors. *Computers in Biology and Medicine, 38*(4), 484–489.

Ganten, M.-K., Weber, T. F., von Tengg-Kobligk, H., Böckler, D., Stiller, W., Geisbüsch, P., et al. (2009). Motion characterization of aortic wall and intimal flap by ECG-gated CT in patients with chronic B-dissection. *European Journal of Radiology, 72*(1), 146–153.

Gao, F., Guo, Z., Sakamoto, M., & Matsuzawa, T. (2006). Fluid-structure Interaction within a Layered Aortic Arch Model. *Journal of Biological Physics, 32*(5), 435–454.

Gijsen, F., van de Vosse, F., & Janssen, J. (1999). The influence of the non-Newtonian properties of blood on the flow in large arteries: Steady flow in a carotid bifurcation model. *Journal of Biomechanics, 32*, 601–608.

Glower, D. D., Speier, R. H., White, W. D., Smith, L. R., Rankin, J. S., & Wolfe, W. G. (1991). Management and long-term outcome of aortic dissection. *Annals of Surgery, 214*(1), 31–41.

Grabenwoger, M., Alfonso, F., Bachet, J., Bonser, R., Czerny, M., Eggebrecht, H., et al. (2012). Thoracic Endovascular Aortic Repair (TEVAR) for the treatment of aortic diseases: A position statement from the European Association for Cardio-Thoracic Surgery (EACTS) and the European Society of Cardiology (ESC), in collaboration with the European Association of Percutaneous Cardiovascular interventions (EAPCI). *European Heart Journal, 33*, 1558–1563.

Grinberg, L., Anor, T., Madsen, J. R., Yakhot, A., & Karniadakis, G. E. (2009). Large-scale simulation of the human arterial tree. *Clinical and Experimental Pharmacology and Physiology, 36*(2), 194–205.

Gysi, J., Schaffner, T., Mohacsi, P., Aeschbacher, B., Althaus, U., & Carrel, T. (1997). Early and late outcome of operated and non-operated acute dissection of the descending aorta. *European Journal of Cardio-Thoracic Surgery, 11*(6), 1163–1170.

Hagan, P. G., Nienaber, C. A., Isselbacher, E. M., Bruckman, D., Karavite, D. J., Russman, P. L., et al. (2000). The international registry of acute aortic dissection (IRAD). *JAMA: The Journal of the American Medical Association, 283*(7), 897–903.

Hanna, J. M., Andersen, N. D., Ganapathi, A. M., McCann, R. L., & Hughes, G. C. (2014). Five-year results for endovascular repair of acute complicated type B aortic dissection. *Journal of Vascular Surgery, 59*(1), 96–106.

Hirst, A. E. J., Johns, V. J. J., & Kime, S. W. J. (1958). Dissecting aneurysm of the aorta: A review of 505 cases. *Medicine, 37*(3), 217–279.

Hou, G., Tsagakis, K., Wendt, D., Stuhle, S., Jakob, H., & Kowalczyk, W. (2010). Three-phase numerical simulation of blood flow in the ascending aorta with dissection. In *Proceedings of the 5th European Conference on Computational Fluid Dynamics (ECCOMAS CFD '10)*.

JCS Joint Working Group. (2013). Guidelines for diagnosis and treatment of aortic aneurysm and aortic dissection (JCS 2011). *Circulation Journal, 77*(3), 789–828.

Johnston, B. M., Johnston, P. R., Corney, S., & Kilpatrick, D. (2004). Non-Newtonian blood flow in human right coronary arteries: Steady state simulations. *Journal of Biomechanics, 37*(5), 709–720.

Jones, D. W., Goodney, P. P., Nolan, B. W., Brooke, B. S., Fillinger, M. F., Powell, R. J., et al. (2014). National trends in utilization, mortality, and survival after repair of type B aortic dissection in the Medicare population. *Journal of Vascular Surgery, 60(1)*(1), 11–19.e1.

Jung, J., & Hassanein, A. (2008). Three-phase CFD analytical modeling of blood flow. *Medical Engineering and Physics, 30*(1), 91–103.

Juvonen, T., Ergin, M. A., Galla, J. D., Lansman, S. L., McCullough, J. N., Nguyen, K., et al. (1999). Risk factors for rupture of chronic type B dissections. *The Journal of Thoracic and Cardiovascular Surgery, 117*(4), 776–786.

Karmonik, C., Bismuth, J., Davies, M. G., Shah, D. J., Younes, H. K., & Lumsden, A. B. (2011a). A computational fluid dynamics study pre- and post-stent graft placement in an acute type B aortic dissection. *Vascular and Endovascular Surgery, 45*(2), 157–164.

References

Karmonik, C., Duran, C., Shah, D. J., Anaya-Ayala, J. E., Davies, M. G., Lumsden, A. B., et al. (2012a). Preliminary findings in quantification of changes in septal motion during follow-up of type B aortic dissections. *Journal of Vacscular Surgery,55*(5), 1419–1426.e1.

Karmonik, C., Bismuth, J. X., Davies, M. G., & Lumsden, A. B. (2008). Computational hemodynamics in the human aorta: A computational fluid dynamics study of three cases with patient-specific geometries and inflow rates. *Technology and Health Care, 16*(5), 343–354.

Karmonik, C., Bismuth, J., Shah, D. J., Davies, M. G., Purdy, D., & Lumsden, A. B. (2011b). Computational study of haemodynamic effects of entry- and exit-tear coverage in a DeBakey type III aortic dissection: technical report. *European Journal of Vascular and Endovascular Surgery, 42*(2), 172–177.

Karmonik, C., Partovi, S., Müller-Eschner, M., Bismuth, J., Davies, M. G., Shah, D. J., et al. (2012c). Longitudinal computational fluid dynamics study of aneurysmal dilatation in a chronic DeBakey type III aortic dissection. *Journal of Vacscular Surgery, 56*(260–263), e1.

Khanafer, K., & Berguer, R. (2009). Fluid-structure interaction analysis of turbulent pulsatile flow within a layered aortic wall as related to aortic dissection. *Journal of Biomechanics,42*, 2642–2648.

Khan, I. A., & Nair, C. K. (2002). Clinical, diagnostic, and management perspectives of aortic dissection. *Chest Journal, 122*(1), 311–328.

Kim, H. J., Vignon-Clementel, I. E., Figueroa, C. A., LaDisa, J. F., Jansen, K. E., Feinstein, J. A., et al. (2009). On coupling a lumped parameter heart model and a three-dimensional finite element aorta model. *Annals of Biomedical Engineering, 37*(11), 2153–2169.

Kinney-Ham, L., Nguyen, H. B., Steele, R., & Walters, E. (2011). Acute aortic dissection in third trimester pregnancy without risk factors. *Western Journal of Emergency Medicine, 12*(4), 571–574.

Kohli, E., Jwayyed, S., Giorgio, G., & Bhalla, M. C. (2013). Acute type A aortic dissection in a 36-week pregnant patient. *Case Reports in Emergency Medicine, 5*, 1–3.

Kromhout, D. (2001). Epidemiology of cardiovascular diseases in Europe. *Public Health Nutrition, 4*(2B), 441–475.

Layton, K. F., Kallmes, D. F., Cloft, H. J., Lindell, E. P., & Cox, V. S. (2006). Bovine aortic arch variant in humans: Clarification of a common misnomer. *American Journal of Neuroradiology, 27*(7), 1541–1542.

Levick, J. (2009). *An indtroduction to cardiovascular physiology* (5th ed.). Hodder Arnold.

Levinson, D. C., Edmaedes, D. T., & Griffith, G. C. (1950). Dissecting aneurysm of the aorta; its clinical, electrocardiographic and laboratory features; a report of 58 autopsied cases. *Circulation, 1*(3), 360–387.

Litchford, B., Okies, J. E., Sugimura, S., & Starr, A. (1976). Acute aortic dissection from cross-clamp injury. *The Journal of Thoracic and Cardiovascular Surgery, 72*(5), 709–713.

Malone, C. D., Urbania, T. H., Crook, S. E. S., & Hope, M. D. (2012). Bovine aortic arch: A novel association with thoracic aortic dilation. *Clinical Radiology, 67*(1), 28–31.

Mani, K., Clough, R. E., Lyons, O. T. A., Bell, R. E., Carrell, T. W., Zayed, H. A., et al. (2012). Predictors of outcome after endovascular repair for chronic type B dissection. *European Journal of Vascular and Endovascular Surgery, 43*(4), 386–391.

Marzo, A., Singh, P., Reymond, P., Stergiopulos, N., Patel, U., & Hose, R. (2009). Influence of inlet boundary conditions on the local haemodynamics of intracranial aneurysms. *Computer Methods in Biomechanics and Biomedical Engineering, 12*(4), 431–444.

Masuda, Y., Yamada, Z., Morooka, N., Watanabe, S. & Inagaki, Y. (1991). Prognosis of patients with medically treated aortic dissections. *Circulation,84*(5 Suppl), III7–13.

Masuda, Y., Takanashi, K., Takasu, J., & Watanabe, S. (1996). Natural history and prognosis of medical treatment for the patients with aortic dissections. *Nihon Geka Gakkai Zasshi, 97*(10), 890–893.

Menter, F. R. (1992). Performance of popular turbulence model for attached and separated adverse pressure gradient flows. *AIAA Journal, 30*(8), 2066–2072.

Menter, F. R., Langtry, R., & Völker, S. (2006). Transition modelling for general purpose CFD codes. *Flow, Turbulence and Combustion, 77*(1–4), 277–303.

Meszaros, I., Morocz, J., Szlavi, J., Schmidt, J., Tornoci, L., Nagy, L., et al. (2000). Epidemiology and clinicopathology of aortic dissection—A population-based longitudinal study over 27 years. *Chest Journal, 117*(5), 1271–1278.

Middleman, S. (1972). *Transport phenomena in the cardiovascular system*. Wiley.

Midulla, M., Moreno, R., Baali, A., Chau, M., Negre-Salvayre, A., Nicoud, F., et al. (2012). Haemodynamic imaging of thoracic stent-grafts by computational fluid dynamics (CFD): presentation of a patient-specific method combining magnetic resonance imaging and numerical simulations. *European Radiology, 22*(10), 2094–2102.

Mligiliche, N. L., & Isaac, N. D. (2009). A three branches aortic arch variant with a bi-carotid trunk and a retro-esophageal right subclavian artery. *International Journal of Anatomical Variations, 2*, 11–14.

Moin, P., & Mahesh, K. (1998). Direct numerical simulation: a tool in turbulence research. *Annual Review of Fluid Mechanics, 30*(1), 539–578.

Morbiducci, U., Ponzini, R., Gallo, D., Bignardi, C., & Rizzo, G. (2013). Inflow boundary conditions for image-based computational hemodynamics Impact of idealized versus measured velocity profiles in the human aorta. *Journal of Biomechanics, 46*(1), 102–109.

Mossop, P. J., McLachlan, C. S., Amukotuwa, S. A., & Nixon, I. K. (2005). Staged endovascular treatment for complicated type B aortic dissection. *Nature Clinical Practice Cardiovascular Medicine, 2*(6), 316–321.

Munson, B., Young, D. & Okiishi, T. (1994). *Fundamentals of fluid mechanics* (2nd ed.). Wiley.

Ngan, K., Hsueh, C., Hsieh, H. C., & Ueng, S. (2006). Aortic dissection in a young patient without any predisposing factors. *Chang Gung Medical Journal, 29*(4), 419.

Nguyen, C. T., Hall, C. S., & Wickline, S. A. (2002). Characterization of aortic microstructure with ultrasound: implications for mechanisms of aortic function and dissection. *IEEE Transactions on Ultrasonics, Ferroelectrics and Frequency Control, 49*(11), 1561–1571.

Nienaber, C. A., Kische, S., Rousseau, H., Eggebrecht, H., Rehders, T. C., Kundt, G., et al. (2013). Endovascular repair of type B aortic dissection: long-term results of the randomized investigation of stent grafts in aortic dissection trial. *Circulation: Cardiovascular Interventions,6*(4), 407–416.

Nienaber, C., Fattori, R., Lund, G., Dieckmann, C., Wolf, W., von Kodolitsch, Y., et al. (1999). Non-surgical reconstruction of thoracic aortic dissection by stent-graft placement. *The New England Journal of Medicine, 340*, 1539–1545.

Nienaber, C. A. (2004). Gender-related differences in acute aortic dissection. *Circulation, 109*(24), 3014–3021.

Nienaber, C. A., Zannetti, S., Barbieri, B., Kische, S., Schareck, W., Rehders, T. C., et al. (2005). INvestigation of STEnt grafts in patients with type B Aortic dissection: Design of the INSTEAD trial-a prospective, multicenter. *European randomized trial. American Heart Journal, 149*(4), 592–599.

Nienaber, C. A., Rousseau, H., Eggebrecht, H., Kische, S., Fattori, R., Rehders, T. C., et al. (2009). Randomized comparison of strategies for type B aortic dissection: The INvestigation of STEnt Grafts in Aortic Dissection (INSTEAD) Trial. *Circulation, 120*(25), 2519–2528.

Nienaber, C. A. (2011). Influence and critique of the INSTEAD trial (TEVAR Versus Medical Treatment for Uncomplicated Type B Aortic Dissection). *Seminars in Vascular Surgery, 24*(3), 167–171.

Nienaber, C. A., Divchev, D., Palisch, H., Clough, R. E., & Richartz, B. (2014). Early and late management of type B aortic dissection. *British Heart Journal, 100*(19), 1491–1497.

Nienaber, C. A., & Eagle, K. A. (2003). Aortic dissection: new frontiers in diagnosis and management: Part II: therapeutic management and follow-up. *Circulation, 108*(6), 772–778.

Olufsen, M. S., Peskin, C. S., Kim, W. Y., Pedersen, E. M., Nadim, A., & Larsen, J. (2000). Numerical simulation and experimental validation of blood flow in arteries with structured-tree outflow conditions. *Annals of Biomedical Engineering, 28*(11), 1281–1299.

Parker, J. D., & Golledge, J. (2008). Outcome of endovascular treatment of acute type B aortic dissection. *The Annals of Thoracic Surgery, 86*(5), 1707–1712.

Parsa, C. J., Daneshmand, M. A., Lima, B., Balsara, K., McCann, R. L., & Hughes, G. C. (2010). Utility of remote wireless pressure sensing for endovascular leak detection after endovascular thoracic aneurysm repair. *ATS, 89*(2), 446–452.

Parsa, C. J., Williams, J. B., Bhattacharya, S. D., Wolfe, W. G., Daneshmand, M. A., McCann, R. L., et al. (2011). Midterm results with thoracic endovascular aortic repair for chronic type B aortic dissection with associated aneurysm. *The Journal of Thoracic and Cardiovascular Surgery, 141*(2), 322–327.

Patel, Y. D. (1986). Rupture of an aortic dissection into the pericardium. *CardioVascular and Interventional Radiology, 9*(4), 222–224.

Patel, H. J., Williams, D. M., Meekov, M., Dasika, N. L., Upchurch, G. R, Jr., & Deeb, G. M. (2009). Long-term results of percutaneous management of malperfusion in acute type B aortic dissection: Implications for thoracic aortic endovascular repair. *The Journal of Thoracic and Cardiovascular Surgery, 138*(2), 300–308.

Pedley, T. J. (1980). *The fluid mechanics of large blood vessels*. Cambridge Monographs on Mechanics and Applied Mathematics.: Cambridge University Press.

Perron, A. D., & Gibbs, M. (1997). Thoracic aortic dissection secondary to crack cocaine ingestion. *The American Journal of Emergency Medicine, 15*(5), 507–509.

Poelma, C., Watton, P. N., & Ventikos, Y. (2015). Transitional flow in aneurysms and the computation of haemodynamic parameters. *Journal of The Royal Society Interface, 12*(105), 20141394.

Popel, A., & Johnson, P. (2005). Microcirculation and hemorheology. *Annual Review of Fluid Mechanics, 37*, 43–69.

Pyeritz, R. E. (2000). The Marfan Syndrome. *Annual Review of Medicine, 51*, 481–510.

Qiao, A., Yin, W., & Chu, B. (2014). Numerical simulation of fluid-structure interaction in bypassed DeBakey III aortic dissection. *Computer Methods in Biomechanics and Biomedical Engineering, 18*(11), 1173–1180.

Qin, Y.-L., Deng, G., Li, T.-X., Wang, W., & Teng, G.-J. (2013). Treatment of acute type-B aortic dissectionthoracic endovascular aortic repair or medical management alone? *JACC: Cardiovascular Interventions, 6*(2), 185–191.

Quemada, D. (1977). Rheology of concentrated disperse systems and minimum energy dissipation principle I. Viscosity-concentration relationship. *Rheologica Acta, 16*, 82–94.

Quemada, D. (1978a). Rheology of concentrated disperse systems II. A model for non-newtonian shear viscosity in steady flows. *Rheologica Acta, 17*, 632–642.

Quemada, D. (1978b). Rheology of concentrated disperse systems III. General features of the proposed non-newtonian model. Comparison with experimental data. *Rheologica Acta, 17*, 643–653.

Rajagopal, K., Bridges, C., & Rajagopal, K. R. (2007). Towards an understanding of the mechanics underlying aortic dissection. *Biomechanics and Modeling in Mechanobiology, 6*(5), 345–359.

Razavi, A., Shirani, E., & Sadeghi, M. R. (2011). Numerical simulation of blood pulsatile flow in a stenosed carotid artery using different rheological models. *Journal of Biomechanics, 44*(11), 2021–2030.

Roberts, W. (1981). Aortic dissection: Anatomy, consequences, and causes. *American Heart Journal, 101*(2), 195–214.

Rudenick, P. A., Bijnens, B. H., Garcia-Dorado, D., & Evangelista, A. (2013). An in vitro phantom study on the influence of tear size and configuration on the hemodynamics of the lumina in chronic type B aortic dissections. *Journal of Vascular Surgery,57*(2), 464–474.e5.

Rudenick, P., Bordone, M., Bijnens, B., Soudah, E., Oñate, E., Garcia-Dorado, D., et al. (2010). A multi-method approach towards understanding the pathophysiology of aortic dissections—The complementary role of in-silico, in-vitro and in-vivo information. In*Statistical Atlases and Computational Models of the Heart* (pp. 114–123). Springer.

Sakai, L. Y., Keene, D. R., & Engvall, E. (1986). Fibrillin, a new 350-kD glycoprotein, is a component of extracellular microfibrils. *The Journal of Cell Biology, 103*(6), 2499–2509.

Sbarzaglia, P., Lovato, L., Buttazzi, K., Russo, V., Renzulli, M., Palombara, C., et al. (2006). Interventional techniques in the treatment of aortic dissection. *La Radiologia Medica, 111*(4), 585–596.

Scali, S. T., Feezor, R. J., Chang, C. K., Stone, D. H., Hess, P. J., Martin, T. D., et al. (2013a). Efficacy of thoracic endovascular stent repair for chronic type B aortic dissection with aneurysmal degeneration. *Journal of Vascular Surgery,58*(1), 10–7.e1.

Schor, J. S., Yerlioglu, M. E., Galla, J. D., Lansman, S. L., Ergin, M. A., & Griepp, R. B. (1996). Selective management of acute type B aortic dissection: Long-term follow-up. *Annals of Thoracic Surgery, 61*(5), 1339–1341.

Shahcheraghi, N., Dwyer, H. A., Cheer, A. Y., Barakat, A. I., & Rutaganira, T. (2002). Unsteady and three-dimensional simulation of blood flow in the human aortic arch. *Journal of Biomechanical Engineering, 124*(4), 378.

Shaw, R. S. (1955). Acute dissecting aortic aneurysm; treatment by fenestration of the internal wall of the aneurysm. *New England Journal of Medicine, 253*(8), 331–333.

Shi, Y., Lawford, P., & Hose, R. (2011). Review of Zero-D and 1-D models of blood flow in the cardiovascular system. *Biomedical Engineering Online, 10*(1), 33.

Shiran, H., Odegaard, J., Berry, G., Miller, D. C., Fischbein, M., & Liang, D. (2014). Aortic wall thickness: An independent risk factor for aortic dissection? *The Journal of Heart Valve Disease, 23*(1), 17–24.

Soudah, E., Rudenick, P., Bordone, M., Bijnens, B., Garcia-Dorado, D., Evangelista, A., et al. (2015). Validation of numerical flow simulations against in vitro phantom measurements in different type B aortic dissection scenarios. *Computer Methods in Biomechanics and Biomedical Engineering, 18*(8), 805–815.

Soulis, J. V., Giannoglou, G. D., Chatzizisis, Y. S., Seralidou, K. V., Parcharidis, G. E., & Louridas, G. E. (2008). Non-Newtonian models for molecular viscosity and wall shear stress in a 3D reconstructed human left coronary artery. *Medical Engineering & Physics, 30*(1), 9–19.

Spittell, P. C., Spittell, J. A. J., Joyce, J. W., Tajik, A. J., Edwards, W. D., Schaff, H. V., et al. (1993). Clinical features and differential diagnosis of aortic dissection: Experience with 236 cases (1980 through 1990). *Mayo Clinic Proceedings, 68*(7), 642–651.

Sughimoto, K., Takahara, Y., Mogi, K., Yamazaki, K., Tsubota, K., Liang, F., et al. (2013). Blood flow dynamic improvement with aneurysm repair detected by a patient-specific model of multiple aortic aneurysms. *Heart and Vessels, 29*(3), 404–412.

Suh, G.-Y., Les, A. S., Tenforde, A. S., Shadden, S. C., Spilker, R. L., Yeung, J. J., et al. (2010). Quantification of particle residence time in abdominal aortic aneurysms using magnetic resonance imaging and computational fluid dynamics. *Annals of Biomedical Engineering, 39*(2), 864–883.

Svensson, L. G., Crawford, E. S., Hess, K. R., Coselli, J. S. & Safi, H. J. (1990). Dissection of the aorta and dissecting aortic aneurysms. Improving early and long-term surgical results. *Circulation,82*(5 Suppl), IV24–38.

Svensson, L. G., Labib, S. B., Eisenhauer, A. C., & Butterly, J. R. (1999). Intimal tear without hematoma: an important variant of aortic dissection that can elude current imaging techniques. *Circulation, 99*(10), 1331–1336.

Svensson, L. G., Kouchoukos, N. T., Miller, D. C., Bavaria, J. E., Coselli, J. S., Curi, M. A., et al. (2008). Expert consensus document on the treatment of descending thoracic aortic disease using endovascular Stent-Grafts. *The Annals of Thoracic Surgery, 85*(1), S1–S41.

Svensson, L. G., & Crawford, E. S. (1992). Aortic dissection and aortic aneurysm surgery: Clinical observations, experimental investigations, and statistical analyses. Part II. *Current Problems in Surgery, 29*(12), 913–1057.

Szeto, W. Y., McGarvey, M., Pochettino, A., Moser, G. W., Hoboken, A., Cornelius, K., et al. (2008). Results of a new surgical paradigm: Endovascular repair for acute complicated type B aortic dissection. *The Annals of Thoracic Surgery, 86*(1), 87–94.

Takahashi, Y., Tsutsumi, Y., Monta, O., Kohshi, K., Sakamoto, T., & Ohashi, H. (2008). Acute onset of paraplegia after repair of abdominal aortic aneurysm in a patient with acute type B aortic dissection. *Interactive Cardiovascular and Thoracic Surgery, 8*(2), 240–242.

References

Tan, F. P. P., Soloperto, G., Bashford, S., Wood, N. B., Thom, S., Hughes, A., et al. (2008). Analysis of flow disturbance in a stenosed carotid artery bifurcation using two-equation transitional and turbulence models. *Journal of Biomechanical Engineering, 130*(6), 061008.

Tan, F. P. P., Borghi, A., Mohiaddin, R. H., Wood, N. B., Thom, S., & Xu, X. Y. (2009). Analysis of flow patterns in a patient-specific thoracic aortic aneurysm model. *Computers and Structures, 87*(11–12), 680–690.

Tang, A. Y. S., Fan, Y., Cheng, S. W. K., & Chow, K. W. (2012). Biomechanical factors influencing type B thoracic aortic dissection: Computational fluid dynamics study. *Engineering Applications of Computational Fluid Mechanics, 6*(4), 622–632.

Taylor, C., Draney, M., Ku, J., Parker, D., Steele, B., Wang, K., et al. (1999). Predictive medicine: Computational techniques in therapeutic decision-making. *Computer Aided Surgery, 4*, 231–247.

Taylor, C. A., & Draney, M. T. (2004). Experimental and computational methods in cardiovascular fluid mechanics. *Annual Review of Fluid Mechanics, 36*(1), 197–231.

Tortora, G. & Grabowski, S. R. (2001), *Introduction to the human body* (5th ed.). Wiley.

Trimarchi, S. (2006). Role and results of surgery in acute type B aortic dissection: Insights from the International Registry of Acute Aortic Dissection (IRAD). *Circulation, 114*(1), 357–364.

Tsai, T. T., Evangelista, A., Nienaber, C. A., Myrmel, T., Meinhardt, G., Cooper, J. V., et al. (2007). Partial thrombosis of the false lumen in patients with acute type B aortic dissection. *New England Journal of Medicine, 357*(4), 349–359.

Tsai, T. T., Fattori, R., Trimarchi, S., Isselbacher, E., Myrmel, T., Evangelista, A., et al. (2006). Long-term survival in patients presenting with type B acute aortic dissection: Insights from the International Registry of Acute Aortic Dissection. *Circulation, 114*(21), 2226–2231.

Tsai, T. T., Schlicht, M. S., Khanafer, K., Bull, J. L., Valassis, D. T., Williams, D. M., et al. (2008). Tear size and location impacts false lumen pressure in an ex vivo model of chronic type B aortic dissection. *Journal of Vascular Surgery, 47*(4), 844–851.

Tsai, T. T., Trimarchi, S., & Nienaber, C. A. (2009). Acute aortic dissection: Perspectives from the International Registry of Acute Aortic Dissection (IRAD). *European Journal of Vascular and Endovascular Surgery, 37*(2), 149–159.

Tse, K. M., Chiu, P., Lee, H. P., & Ho, P. (2011). Investigation of hemodynamics in the development of dissecting aneurysm within patient-specific dissecting aneurismal aortas using computational fluid dynamics (CFD) simulations. *Journal of Biomechanics, 44*(5), 827–836.

Umaña, J. P., Lai, D. T., Mitchell, R. S., Moore, K. A., Rodriguez, F., Robbins, R. C., et al. (2002). Is medical therapy still the optimal treatment strategy for patients with acute type B aortic dissections? *The Journal of Thoracic and Cardiovascular Surgery, 124*(5), 896–910.

Valen-Sendstad, K., Mardal, K.-A., Mortensen, M., Reif, B. A. P., & Langtangen, H. P. (2011). Direct numerical simulation of transitional flow in a patient-specific intracranial aneurysm. *Journal of Biomechanics, 44*(16), 2826–2832.

Valen-Sendstad, K., Mardal, K.-A., & Steinman, D. A. (2013). High-resolution CFD detects high-frequency velocity fluctuations in bifurcation, but not sidewall, aneurysms. *Journal of Biomechanics, 46*(2), 402–407.

Valen-Sendstad, K., Piccinelli, M., & Steinman, D. A. (2014). High-resolution computational fluid dynamics detects flow instabilities in the carotid siphon: Implications for aneurysm initiation and rupture? *Journal of Biomechanics, 47*(12), 3210–3216.

Versteeg, H., & Malalasekera, W. (2007). *An introduction to computational fluid dynamics* (2nd ed.). Prentice Hall.

Weigang, E., Nienaber, C. A., Rehders, T. C., Ince, H., Vahl, C.-F., & Beyersdorf, F. (2008a). Management of patients with aortic dissection. *Deutsches Ärzteblatt international, 105*(38), 639–645.

Wilson, S. K., & Hutchins, G. M. (1982). Aortic dissecting aneurysms: causative factors in 204 subjects. *Archives of Pathology & Laboratory Medicine, 106*(4), 175–180.

Yacoub, M., ElGuindy, A., Afifi, A., Yacoub, L., & Wright, G. (2014). Taking cardiac surgery to the people. *Journal of Cardiovascular Translational Research, 7*(9), 797–802.

Yamashiro, S., Kuniyoshi, Y., Miyagi, K., Uezu, T., Arakaki, K., & Koja, K. (2003). Acute postoperative paraplegia complicating with emergency graft replacement of the ascending aorta for the type a dissection. *Annals of Thoracic and Cardiovascular Surgery, 9*(5), 330–333.

Yang, S., Li, X., Chao, B., Wu, L., Cheng, Z., Duan, Y., et al. (2014). Abdominal aortic intimal flap motion characterization in acute aortic dissection: Assessed with retrospective ECG-gated thoracoabdominal aorta dual-source CT angiography. *PLoS ONE, 9*(2), e87664.

Chapter 2
Computational Methods for Patient-Specific Modelling

Patient-specific haemodynamic modelling using computational fluid dynamics approaches is a multi-stage process. Firstly, a model of the geometry of interest is created and discretised. The governing equations for the fluid must then be solved for each discretised element, and the interfaces of the domain should be treated appropriately. This chapter will describe the basic equations solved using CFD and their numerical treatment. Subsequently, the stages required in imaging the patient and converting the images into a 3D geometry will be given, followed by a brief description of discretisation (meshing). The mathematics behind lumped-parameter modelling to develop dynamic BCs will be described and a comparison with alternative BCs will be made. Finally, a brief introduction to the relevant aspects of solid modelling will be provided.

2.1 Computational Fluid Dynamics

The motion of a fluid can be described by three conservation principles: mass, momentum and energy. Mass conservation states that for a system, mass cannot be created or destroyed (Munson et al. 1994). Conservation of momentum is described by Newton's second law and states that the summation of the forces applied on a fluid particle is equal to the rate of change of momentum (Fung 1993). These forces can be body forces such as gravity, centrifugal and electromagnetic forces, or they can be surface forces such as pressure and viscous forces (White 2011). The conservation of energy states that energy cannot be destroyed or created, but only converted from one form into another. In the context of a finite volume, the rate of change of energy is thus the difference between the energy going into and out of the volume. In the aorta, heat transfer is not a major component of the fluid dynamics, and modelling

---

A preliminary version of the 3D-0D coupling approach developed in this Chapter was used in a study on haemodynamics in an arterio-venous fistula (Decorato et al. 2014).

© Springer International Publishing AG 2018
M. Alimohammadi, *Aortic Dissection: Simulation Tools for Disease Management and Understanding*, Springer Theses, https://doi.org/10.1007/978-3-319-56327-5_2

the system as isothermal is sufficient. As such, the energy equation does not need to be solved to describe the system, and will not be discussed further.

2.1.1 Governing Equations

The fluid dynamics of an incompressible, Newtonian fluid undergoing laminar flow in an isothermal environment can be described can be mathematically described by the continuity equation and the three linear momentum equations known as the Navier-Stokes equations (Versteeg and Malalasekera 2007; White 2011). The flow of liquids at body temperature ($\approx 37°$) can be considered incompressible, and thus the following equations are stated for constant density (Fung 1997). Additionally, gravitational body forces are assumed to be negligible. The continuity equation can be written in the form:

$$\nabla \mathbf{U} = 0 \tag{2.1}$$

The Navier-Stokes equations for incompressible, laminar flow are given by

$$\rho \left(\frac{\partial \mathbf{U}}{\partial t} + \mathbf{U}.\nabla \mathbf{U} \right) = -\nabla P + \mu \nabla^2 \mathbf{U} \tag{2.2}$$

where P is the pressure and μ is the dynamic viscosity. The Navier-Stokes equations are non-linear partial differential equations and thus can only be solved for a few simplified problems. One such example is Hagen-Poiseuille flow, commonly used for blood flow analysis, which describes the steady, fully developed, axisymmetric flow of an incompressible, laminar, Newtonian fluid through a cylinder of constant diameter D of length L. Thus the velocity is only a function of the radius and the only non-zero pressure term is in the axial direction. Although this is never the case in the vasculature, Poiseuille flow is often used as a benchmark for various haemodynamic parameters. The 'Hagen-Poiseuille law' states the pressure drop in terms of the flow rate according to

$$\Delta P = \frac{128 \mu Q L}{\pi D^4} = RQ \tag{2.3}$$

where $R = 128\mu L/\pi D^4$ is the hydrodynamic resistance. In blood vessels, one or more of the assumptions made in the derivation of Eq. 2.3 is not appropriate. Therefore, in order to simulate cardiovascular flows, numerical solutions are required. This is carried out by discretising the fluid domain into small elements, and finding the solution that has the smallest error for Eqs. 2.1 and 2.2 for every element in the domain.

2.1.2 Numerical Implementation

Numerical implementation of CFD is a well-developed field, and several textbooks are available on the topic (Versteeg and Malalasekera 2007; Wesseling 2000). Furthermore, a range of commercial packages are available, such STAR-CD (CD-adapco, Melville, NY, USA), Phoenics (Concentration Heat and Momentum, London, UK), Fluent and CFX (ANSYS, Canonsburg, PA, USA). Each package uses different methods for solving the equations. In the present study, the software ANSYS CFX will be used. ANSYS provides detailed explanation of the numerical methods utilised by CFX in the manuals for the software. Details relating to the specific methods used for the present simulations are described briefly in the following paragraphs, summarised from the CFX manual (ANSYS 2011).

CFX uses the finite volume method to solve the Navier-Stokes equations numerically.

CFX constructs finite volumes about the nodes in a mesh, with surfaces connecting the centre of connected elements to the centre of the element edges, as shown in Fig. 2.1a. Integration points are created at the centre of each surface element (Fig. 2.1b). In order to solve the discretised Navier-Stokes equations, Gauss divergence theorem is applied, which converts volume integrals into surface integrals, and so the large number of surfaces increases the accuracy per control volume. As compared to a cell-centred method, wherein the mesh elements from the imported mesh are used directly, the vertex-centred method used by CFX uses fewer control volumes with increased accuracy, and thus increases overall efficiency.

Although finer meshes reduce discretisation errors, they also have slower convergence rates (Versteeg and Malalasekera 2007), and thus for complex geometries involving hundreds of thousands of nodes, the computational cost can be very large. When using iterative solution methods, this occurs because although local discretisation errors are rapidly reduced, errors on longer length scale decay slowly through the grid. To account for this, ANSYS uses an algebraic multigrid approach, wherein solutions are acquired on fine and coarse grids, the latter constructed by analysing subsets of the system matrix containing the equation coefficients. The solution

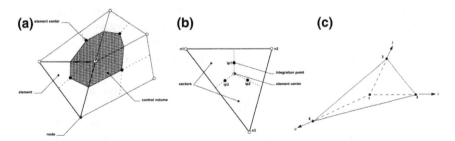

Fig. 2.1 Mesh discretisation in CFX (ANSYS 2011). **a** Control volume definition, **b** Integration points on element edges, **c** tri-linear interpolation for a tetrahedral element

variables are stored at the mesh nodes, as opposed to using a staggered grid layout. However, for many of the terms solved in the discretised Navier-Stokes equations, solutions for the variables are required at the integration points. Shape functions are used to interpolate the values from the surrounding nodes using tri-linear interpolation (Fig. 2.1c).

The $\mathbf{U}.\nabla\mathbf{U}$ term in Eq. 2.2 represents the advective terms: the transport of a property ϕ due to bulk motion of the fluid. For a given node i (defined at the centre of a control volume), the property ϕ_i must be calculated. Central differencing considers equally ϕ_{i-1} and ϕ_{i+1}. However, if the flow is moving from $i-1$ towards $i+1$, then node $i-1$ should have a greater influence on node i. The upwind differencing scheme directly translates ϕ_{i-1} to ϕ_i, which is a robust approach, but relatively inaccurate, and prone to introducing numerical diffusion (Versteeg and Malalasekera 2007). A correction term, scaled by a parameter ψ can be added to ϕ_i to also consider the gradients of ϕ at $i-1$ and $i+1$. With $\psi = 1$, the solution is more accurate, but prone to poor convergence. Thus, ANSYS provides a 'high-resolution' scheme with ψ defined according to boundedness principles (Barth and Jespersen 2012), with ψ calculated for each node so as to optimise accuracy whilst retaining robustness. For all of the results presented in this thesis, the high-resolution scheme was used for the advection terms.

The transient term, $\partial\mathbf{U}/\partial t$, is solved using the second order backward Euler scheme, which is more accurate than the first order scheme. In the first order scheme, the time derivative is estimated according to the difference between the values of ϕ at a given node for the previous and current timesteps, $\partial\phi_i/\partial t = (\phi_i - \phi_{i-1})/\Delta t$. However, this can lead to false diffusion similar to that observed in the upwind differencing scheme. The second order backward Euler scheme estimates the value at the start of the time step based on the previous timestep, plus half the difference between the previous timestep and that preceding it, $\phi_{i_{start}} = \phi_{i-1} + 0.5(\phi_{i-1} - \phi_{i-2})$. The value of ϕ at the end of the timestep is estimated similarly, $\phi_{i_{end}} = \phi_i + 0.5(\phi_i - \phi_{i-1})$, and the derivative is given by $\partial\phi_i/\partial t = (\phi_{i,start} - \phi_{i,end})/\Delta t$. This approach is more accurate than the first order scheme, but is also robust.

Shape functions using tri-linear interpolation (Fig. 2.1c) are used to calculate the derivatives required for the diffusion terms, $\nabla^2 \mathbf{U}$. The pressure gradient term is similarly calculated using shape functions, but with linear-linear interpolation.

As ANSYS uses a non-staggered grid layout, there is a risk of obtaining a 'checkboard' pressure field (Patankar 1980), in which calculation of local gradients yields alternately high and low regions (Versteeg and Malalasekera 2007). The 4th order Rhie-Chow scheme is used to avoid a decoupled pressure field (Rhie and Chow 2012).

Many CFD codes use segregated solvers, in which a pressure is guessed and used to solve the momentum equations, from which a pressure correction term is calculated. This approach typically requires a large number of iterations, and so in order to reduce computational time CFX uses a fully coupled approach, in which the equations for all three velocity components and the pressure are solved simultaneously. The general strategy is as follows. Consider a general set of linear equations, $A\phi = b$, where A is the coefficient matrix, b is the right hand side of the equations and ϕ is the solution

2.1 Computational Fluid Dynamics

vector. An initial solution, ϕ_n is improved according to $\phi_{n+1} = \phi_n + \phi'$. ϕ' is the solution of $A\phi' = r_n$, and r_n is the residual, which can be calculated according to $r_n = b - A\phi_n$. This process is repeated until r is lower than the desired threshold.

2.2 Building the Fluid Domain

The term 'patient-specific' modelling implies that the model contains some elements specific to a given patient. In general as much information as is available is used, but perhaps the most fundamental aspect is to use a realistic morphological representation of the patient's anatomy in the region of interest. Anatomical details vary considerably within the population, but in the context of AD, the variability is far greater. Differences in numbers of tears, locations of tears, partial thromboses, coarctations, aneurysms etc. make each case of AD highly unique. The diagnosis of AD is made based on clinical imaging, typically using either computed tomography (CT) or magnetic resonance imaging (MRI).

2.2.1 Introduction to Clinical Imaging

Radiography (imaging via ionising radiation) was the first in vivo imaging method and was discovered by Wilhelm Roentgen in 1895 (Studwell and Kotton 2011). This discovery led to the development of X-ray imaging. X-rays are used to show the interior of a mammalian body by detecting the different densities of each tissue. Computed tomography uses ionising radiation in a number of slices which can be converted into three-dimensional (3D) organs or tissue structures. The advantages of using CT scans are the high spatial resolution, high penetration depth and clinical translation. However, they have limited sensitivity and a disadvantage of radiation exposure (Studwell and Kotton 2011).

An alternative imaging modality is magnetic resonance imaging (MRI), which captures the geometry of the internal tissue by absorption and emission of electromagnetic energy (electromagnetism). MRI yields an image based on the radiofrequency of the tissue, which is generated by vibrating hydrogen atoms within the tissues. MRI has lower spatial resolution than CT, but has the significant advantages that no radiation exposure occurs and a contrast-agent is not necessary (JCS Joint Working Group 2013) (although contrast-agent may be used in certain MRI modalities). However, due to the more time-consuming nature of MRI, it is not routinely used for patient monitoring and is not recommended for acute aneurysms (JCS Joint Working Group 2013). Additionally, the high cost (Beynon et al. 2012) and scarce resources (Taylor et al. 2005) limits the clinical use of MRI in public healthcare.

Using the electrocardiogram (ECG) as a trigger, it is possible to get time-resolved CT data (Yang et al. 2014; Ganten et al. 2009) from which wall deformation can be calculated, although the spatial resolution remains a limiting factor. Time-resolved data at 2D planes can also be acquired with 2D pcMRI (Karmonik et al. 2012a), which can also provide data on flow rates. A number of researchers have recently reported use of 4D MRI in AD (Francois et al. 2013; Clough et al. 2012), which can yield full time-resolved data on the geometry and velocities throughout the aorta. While this technique is extremely promising, limitations with spatial and temporal resolution make it difficult to accurately measure slow flow in the FL and small intimal tears (Francois et al. 2013). Combining CFD with MRI has potential as a mixed approach to provide further insight on the haemodynamics of AD, whilst validating the computational results (Cheng et al. 2014; Tan et al. 2009; Karmonik et al. 2012).

In the present study, a CT scan from a 54 year old female patient suffering from type-B AD will be used throughout. The only available clinical data for this patient are CT scans in addition to invasive pressure measurements reporting maxima and minima of the pressure at a number of locations within the aorta. These data were acquired as part of the standard clinical protocol. The study was ethically approved (NHS Health Research Authority, ref: 13/EM/0143) and consent obtained.

2.2.2 3D Domain Extraction

There are a variety of commercially available tools for extracting 3D geometries from imaging data, such as MiMiCs (Materialise, Leuven, Belgium) or ScanIP (Simpleware, Exeter, UK). For the present study, ScanIP image processing software is used to extract the 3D fluid domain representing the aorta of the patient. To do this, a combination of different masks, thresholds and smoothing operations were used to generate the final fluid domain.

A stack of 887 DICOM images of a patient suffering from a type-B dissected aorta was imported into the ScanIP software. These images provide a stack of 2D planes with a resolution of 0.7 mm/pixel and a 0.7 mm distance between planes, with the stack starting from the lower jaw and ending just below the iliac bifurcation. The area of interest was selected so as to include a section of the supraaortic branches and a section of the TL distal to the FL, but prior to the iliac bifurcation. The invasive pressure measurements for the patient were acquired at ascending aorta (close to the aortic valve), almost 3 cm along the brachiocephalic trunk and common carotid artery, and in the TL in the distal arch, mid abdominal aorta and distal abdominal aorta, prior to the iliac bifurcation.

The trimmed stack comprised 473 DICOM images, which corresponded to approximately 330 mm (the total length of the dissected aorta including the supraaortic branches). For the sake of efficiency, the entire stack was also cropped in the plane of each image, by removing the areas located outside the patients ribs (Fig. 2.2).

The segmentation methodology used for the present study is based on analysis of greyscale intensities. The imported images are 8-bit, and thus the intensities range between 0 and 255. A semi-automated process was used to identify the vessels of interest.

2.2 Building the Fluid Domain 45

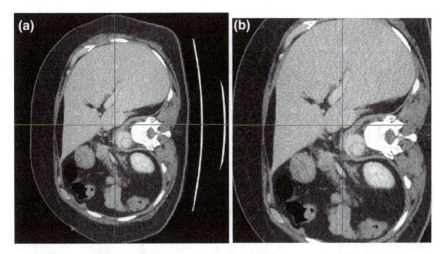

Fig. 2.2 CT scan slices showing **a** the entire domain prior to the cropping stage and **b** the cropped domain in order to reduce the image processing time (only one slice is shown for brevity)

A representative image was selected from the stack (Fig. 2.3a). A line was drawn manually along the vessel lumen from which the software generated a pixel intensity profile (showing the intensity at each point on the line). Upper and lower threshold bounds were selected so as to include most pixels in the lumen. However, in doing so additional regions of similar intensity are selected, such as the spine and ribs, as shown in Fig. 2.3b. However, these hard tissue sections have an identifying feature, which is that the edges are particularly bright. A mask is created by defining a second

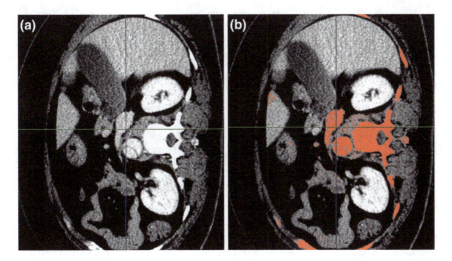

Fig. 2.3 CT scan slices showing **a** a representative image, **b** the same image with the mask superimposed mask in *red*

pixel threshold (>250), which is sufficiently high to only select the hard tissue at the edges. The inner parts of the hard tissue are not selected by this threshold, but are included in the mask using a region growing approach, whereby the bone section is selected and the software automatically fills in the area based on local pixel intensities.

Therefore, to reduce the image processing time even further, a simple methodology is applied that differentiates the hard tissues from the soft tissues that hold the same image properties/intensities when applying a threshold. This results in two sets of separate masks with different thresholds; one for the hard tissues only (spine and ribs) and the other one including the dissected aorta and the hard tissues (Figs. 2.4a, b). As the intensity of the ribs/spine is close to the dissected regions (Fig. 2.4a), one

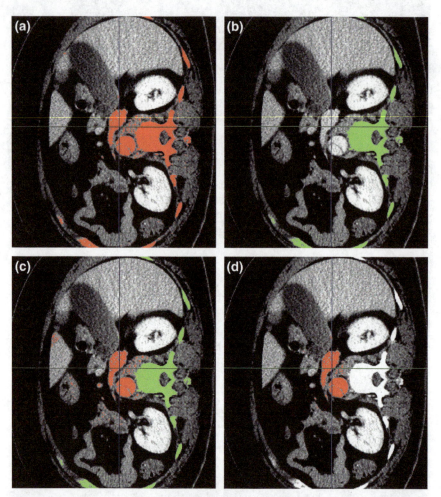

Fig. 2.4 CT scan slices with different thresholds. **a** Dissected aorta and the hard tissues with intensity values above 220 shown in *red*, **b** the hard tissues with intensity values above 250 shown in *green*, **c** both masks on the same image, **d** mask resulting from boolean operation subtracting the hard tissues

2.2 Building the Fluid Domain

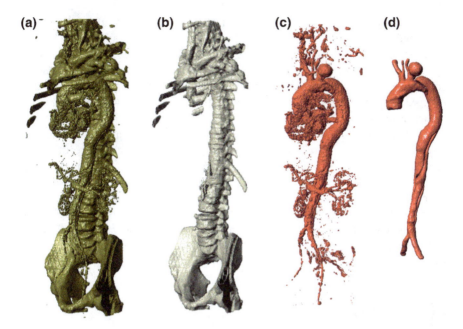

Fig. 2.5 Stages in elimination of the hard tissues. 3D masks of the **a** soft and hard tissue **b** hard tissue only **c** soft tissue only, **d** the smoothed 3D domain

must choose a very high intensity range (>250) that covers the hard tissues without including the soft tissues. The resulting mask is shown in red in Fig. 2.4b.

The settings defined for the representative image are then used to construct 3D masks of the hard tissues, and combined hard and soft tissue (Fig. 2.5a, b). Figure 2.4c shows the 2D projection of both masks on the plane of the representative image. Subtracting the hard tissue mask from the combined hard and soft tissue mask leaves only the soft tissue (Figs. 2.4d and 2.5c), plus a degree of noise due to the varying pixel intensities through the stack of images.

The noise, which comprises part of the heart and kidneys, as well as random voxels must be removed to leave the final geometry. The 'island removal' tool is used to remove isolated 'islands' that are smaller than a selected size. The larger tissue sections such as the heart and kidneys are manually removed by cropping.

Subsequently, smoothing filters are applied to the mask. The smoothing filter aims to reduce any remaining noise that exists in the 3D domain of the dissected aorta using a recursive Gaussian operation. The smoothing filter uses a spatial parameter (Gaussian σ) in mm for all three directions of X, Y and Z and defines how many neighbouring pixels should be accounted for, in order to smooth the mask. The larger this parameter is, the broader the smoothing operation will become, and small structures will be removed. To avoid excessive smoothing, and thus removal of morphological features, only a Gaussian sigma of $\sigma = 0.5$ mm was used (less than one pixel, thus only significantly affecting adjacent pixels). An additional mean filter is applied to further reduce noise. This filter sets each pixel value to the average of its surrounding values. Applying smoothing filters results in a total volume reduction of the entire 3D domain; hence dilation is carried out in conjunction with the smoothing, in order to ensure that the final volume of the 3D domain is not affected by the smoothing operations. The importance of the smoothing operations can be seen when considering the texture of the surface of the aorta in Fig. 2.5c, d. The roughness apparent in Fig. 2.5c is an artefact of the segmentation process which if not smoothed would induce errors in the CFD data.

As described previously, the invasive pressure measurements (only maximum and minimum values) were only available at the ascending aorta, at two of the three supraaortic branches, in the distal abdominal aorta and in two additional locations along the descending aorta. This meant there were no invasive pressure data for the renal arteries or other branches off the dissected aorta (celiac trunk, mesenteric artery). Furthermore, the CT scan resolution is such that it was difficult to accurately extract these small arteries from the present data with sufficient confidence. For both of these reasons, these branches were not included in the model extracted for this study. The final 3D geometry (domain) is shown in Fig. 2.5d.

2.2.3 Geometry

Figure 2.6 shows the final geometry separated into the arch, branches and TL (Fig. 2.6a), and the FL only (Fig. 2.6b). The FL is larger than the TL, and the

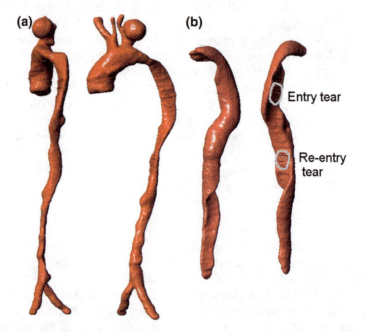

Fig. 2.6 The final 3D geometry prior to the cropping stage. **a** The aortic arch, TL and the iliac bifurcation in two views, **b** the *right* anterior and *left* posterior views of the FL

2.2 Building the Fluid Domain   49

proximal FL is slightly larger than the distal FL (Fig. 2.6b). The locations of entry and re-entry tears are indicated in Fig. 2.6b. The FL and TL are intertwined with the entry and re-entry tears connecting the two lumina together. The TL narrows after the aortic arch (severe coarctation) and continues with a similar cross-sectional area all along the descending aorta. A slight expansion can be seen in TL prior to the iliac bifurcation (Fig. 2.6a).

The reconstructed domain can be seen in Fig. 2.7a. The 3D geometry was imported into CATIA (Dassault Systems, Velizy-Villacoublay, France), wherein the inflow/outflow boundaries were cropped perpendicular to the z-axis (providing a flat surface). The descending aorta was cropped prior to the iliac bifurcation, coincident with the most downstream invasive pressure measurements. The resulting model can be seen in Fig. 2.7b. Blood enters the ascending aorta (AA) and flows into

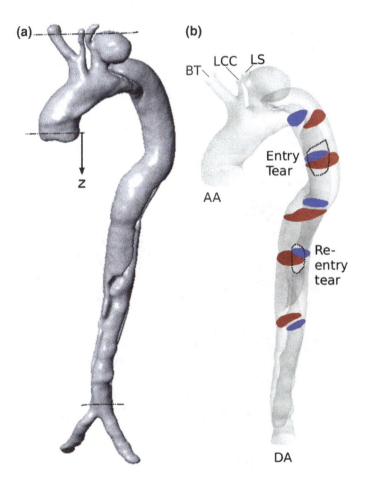

Fig. 2.7 a Patient-specific reconstructed geometry. b The final 3D domain with cropped boundaries (*right*). Selected planes show the TL (*blue*) and FL (*red*) along the descending aorta

the aortic arch. It can be observed that the patient has a 'bovine aortic arch', i.e. the brachiocephalic trunk (BT) and left common carotid artery (LCC) share a common trunk origin (Layton et al. 2006). The second vessel branching off the aortic arch is the left subclavian artery (LS) and a large aneurysm can be seen at the origin of this branch. At the distal arch, a coarctation is present as a result of the proximal expansion of the FL. The entry tear can be observed in the proximal descending aorta, with a roughly elliptical shape, as indicated in Fig. 2.7b. The re-entry tear is located further downstream and is of similar shape to the entry tear, but is smaller in size. The FL extends upwards from the entry tear and downwards from the re-entry tear such that it is present throughout the entire descending aorta.

A number of planes have been added to Fig. 2.7, showing the TL in blue and the FL in red, for visualisation purposes. It can be seen that the cross sectional area of the FL is generally greater than that of the TL; the luminal area of TL is reduced by more than 50% due to the dissection. Figure 2.8 shows the surface area of each boundary, in addition to the largest area section of the aneurysm at the origin of the LS branch. The cross-sectional area of the aneurysm is slightly larger than the AA. The LCC branch has the smallest surface area at its boundary and the AA has the largest surface area of $8.7 \times 10^{-4}\,\mathrm{m}^2$, with an equivalent diameter of approximately 33 mm. Due to the limited resolution of the CT scans, it was not possible to accurately include the visceral arteries in the geometric reconstruction, and they were therefore not included in the model.

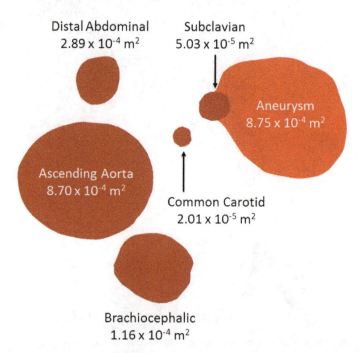

Fig. 2.8 Surface area at each of the domain boundaries and the largest section of the aneurysm

2.3 Meshing

Mesh generation is a procedure of spatial subdivision of the 3D domain into small volumes. These small volumes are referred to as mesh elements. The grid types and guideline sizes are inputs to the mesh generation process, and can be defined for both bodies and separate faces (such as a refinement at a boundary). The generated mesh can be structured or unstructured. Structured meshes use regular patterns based on Cartesian or curvilinear grids. In order to achieve sufficient refinement near key geometrical features, block structured grids can be used, in which key regions are subdivided. However, in the tortuous and morphologically complex features of a dissected aorta, structured meshes are not practicable. Unstructured meshes are better suited to such cases, and allow for easy and computationally efficient refinement of mesh elements where required (Versteeg and Malalasekera 2007).

For the present study, meshing was carried out using ANSYS ICEM. The mesh sizing was controlled with a minimum element size of 1 mm and a seed size of 2.5 mm as initial parameters. Automatic mesh refinement close to morphological features and regions of significant curvature was enabled, and an initial mesh was generated using the octree method. This method generates a coarse grid, then subdivides the elements until almost all mesh elements exceed a predefined minimum quality. Within this stage, prismatic layers (inflation layers), are created along the walls. The prismatic layers consist of quadratic elements (as opposed to tetrahredral elements used in the rest of the mesh) and are structured so as to expand away from the wall. A total of 7 prismatic layers were used, with an expansion ratio of 1.2 (meaning that each layer is 1.2 times thicker than the preceding layer moving towards the centre of the lumen). The use of prismatic layers is an efficient method to refine the mesh near the walls without creating prohibitively large numbers of tetrahedral mesh elements, which would be required in order to accurately resolve the fluid dynamics at the vessel boundary. The generated mesh was post-processed by iterative smoothing in order to ensure a minimum quality for all mesh elements. Finally, the mesh was reprocessed using the Delaunay meshing scheme. Whereas the octree method is more robust, it can result in high gradients in the mesh refinement. This can occur in regions where the algorithm subdivides one volume many times in order to achieve sufficient resolution near a feature, but the adjacent region remains coarse. The Delaunay principle aims to maximise the minimum angle in each mesh element (in this sense a regular tetrahedron is considered to be the ideal element). In doing so, gradients in refinement are smoothed out and a more continuous mesh is generated.

The mesh used for the analysis of the results in Chaps. 3, 4 and 5 contained approximately 250, 000 mesh elements. For each results chapter, a mesh sensitivity analysis was carried out and can be found therein. These analyses ensured that the mesh was sufficiently refined, so that the solution did not change significantly upon further refinement.

2.4 Dynamic Boundary Conditions

The left ventricle fills with blood from the left atrium and the pressure rises to approximately 120 mmHg. When the pressure exceeds that in the aorta, the aortic valve opens and blood rushes from the ventricle, until the pressure in the ventricle is less than in the aorta and the valve closes. During this process the pressure in the ventricle varies from around 10 (Peterson et al. 1978) to 120 mmHg, but the pressure in the aorta remains relatively high throughout (80–120 mmHg). This is a result of the elasticity of the aortic wall, which stores energy during systole and releases it during diastole, ensuring a steady supply of blood to the systemic vasculature. This is commonly known as the Windkessel effect (Shi et al. 2011). In 1889 Otto Frank described the Windkessel model mathematically, and the model can be seen as an analogy with a simple electric circuit in which different physical phenomena are represented by electronic components. This approach is variously known as the hydraulic-electrical analogy, zero-dimensional (0D: as no spatial dimensions are directly involved), lumped-parameter or Windkessel modelling (Fung 1997; Shi et al. 2011).

Lumped parameter models assume a uniform distribution of haemodynamic parameters such as velocity and pressure and the hydraulic-electrical analogy can be used to describe a compartment (or any combination of them) within the vasculature. In this analogy, blood flow and pressure correspond to current and voltage in a circuit respectively. In a circuit, the voltage difference drives the current to flow across the electric impedance of resistive, capacitative and inductive elements. In the vascular analogue, the pressure difference drives the flow across the vasculature, losing energy due to viscous losses in the process (resistance). The passive volume changes in the vasculature due to the vessel wall elasticity are analogous to compliance, and inertial forces can be modelled by inductors.

2.4.1 Analogue Equations

The voltage drop V across a resistor R is given by Ohm's law:

$$V = i_r R \tag{2.4}$$

where i_r is the current. This can be compared to Poiseuille's law, $\Delta P = QR$, where R is the hydrodynamic resistance (see Eq. 2.3). The current across a capacitor is given by

$$i_c = C \frac{dV}{dt} \tag{2.5}$$

2.4 Dynamic Boundary Conditions

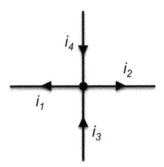

Fig. 2.9 The current entering any node in a circuit is equal to the summation of currents exiting that node

where i_c is the current across the capacitor, V is the voltage and C is the capacitance. This is analogous to stating that the compliance is equal to a volume change for a given pressure change (as volume is analogous to electric charge).

Conservation of mass is equivalent to Kirchoff's current law (KCL), as indicated in Fig. 2.9, which can be written as

$$i_1 + i_2 = i_3 + i_4 \tag{2.6}$$

The simplest lumped parameter model for a section of the vasculature is a resistor. This is only valid if the capacitance is negligible, so is not appropriate for large vessels, but can be used to model the hydrodynamic resistance of, for example, the capillary bed.

2.4.2 Two-element Windkessel Model

For larger vessels, wherein the compliance is not negligible, an RC circuit is the simplest representation, as shown in Fig. 2.10a. The two components are arranged in parallel with a voltage source. This configuration is known as a two-element Windkessel model (WK2). The current entering the circuit is i and the voltage across the circuit is given by V. Using Eqs. 2.4, 2.5 and 2.6 the circuit yields

$$i = C\frac{dV}{dt} + \frac{V}{R} \tag{2.7}$$

Translating Eq. 2.7 into the vascular analogue terms yields

$$Q = C\frac{dP}{dt} + \frac{P}{R} \tag{2.8}$$

as shown in Fig. 2.10b. The WK2 model is very simple, yet is capable of predicting the pressure/flow relationships in the arteries with a reasonable degree of accuracy (Stergiopulos et al. 1999; Westerhof et al. 2008). Using measured pressure or flow

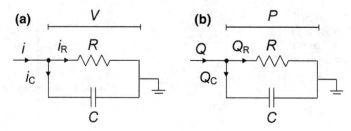

Fig. 2.10 Circuit for the two element Windkessel model

as an input, model parameters can be derived by minimising the difference between the model and measured outputs. These calculated values can be used to provide diagnostic information for clinicians or as parameters for models in research. WK2 models could be used in clinical practice, if the pressure pulse waveform and the peripheral resistance are known, to estimate the total arterial compliance of a system (Levick 2009).

However, the model is based on the assumption that the venous system is a pressure sink (i.e. zero pressure—equivalent to ground in the electrical circuit) and thus it does not yield data on pressure in the venous system. Furthermore, it cannot provide insight into wave travel and reflection, which are important phenomena in large arteries. The wave characteristics result in high frequency responses that cannot be accounted for in the WK2 model, as it only has a single time constant ($\tau_t = RC$).

2.4.3 Three-element Windkessel Model

In order to improve the high-frequency response, an additional component can be added to the circuit in series with the WK2 to represent the characteristic impedance of the vasculature.

The addition of the characteristic impedance (R_1—Fig. 2.11) yields a 3-component RCR circuit known as a WK3 or Westkessel model. Note that this circuit is no longer a direct analogue, as the additional component does not represent a resistance through which the flow must pass before reaching the capacitance (Vignon-Clementel et al.

Fig. 2.11 Circuit for the three element Windkessel model

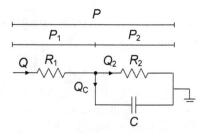

2.4 Dynamic Boundary Conditions

2006). Rather, it mathematically allows for a significant improvement in the modelling of the high frequency characteristics of the system and can, for example, be used to estimate the load on an isolated heart (Stergiopulos et al. 1999). Using a simple resistor for the characteristic impedance may cause some errors in the case of low frequencies, but this is acceptable as the characteristic impedance is only 5–7% of the total peripheral resistance in mammals (Westerhof et al. 2008). It should be noted that during diastole, almost two thirds of the cardiac cycle, both two and three element Windkessel models behave similarly (Westerhof et al. 2008).

For the circuit shown in Fig. 2.11, the relevant equations can be derived as follows, using the fluid analogue terms for clarity. The flow rate into the capacitative element is

$$Q_C = C \frac{dP_2}{dt} \quad (2.9)$$

the flow across R_2 is

$$Q_2 = \frac{P_2}{R_2} \quad (2.10)$$

and continuity (equivalent to KCL) yields

$$Q_C = Q - Q_2 \quad (2.11)$$

Combining Eqs. 2.9, 2.10 and 2.11 yields

$$\frac{dP_2}{dt} = \frac{R_2 Q - P_2}{R_2 C} \quad (2.12)$$

The pressure P_2 is internal to the circuit, and for the present application, the pressure drop across the entire circuit (P) is required, thus P_2 should be eliminated from Eq. 2.12. Based on Kirchoff's voltage law:

$$P_2 = P - R_1 Q \quad (2.13)$$

Substituting into Eq. 2.12 yields

$$\frac{d(P - R_1 Q)}{dt} = \frac{R_2 Q - P + R_1 Q}{R_2 C} \quad (2.14)$$

which simplifies to

$$P = Q(R_1 + R_2) - R_2 C \frac{dP}{dt} + R_1 R_2 C \frac{dQ}{dt} \quad (2.15)$$

In order to solve Eq. 2.15 implicitly within the CFD simulation, the equation must be discretised. This was done using the backward Euler method. Let n be the current timestep and $n - 1$ be the previous timestep, with a timestep length Δt. The pressure derivative is thus calculated according to

$$\frac{dP_n}{dt} = \frac{P_n - P_{n-1}}{\Delta t} \tag{2.16}$$

Substituting Eq. 2.16 into Eq. 2.15 and rearranging for P_n yields

$$P_n = \frac{(R_1 + R_2)Q_n + \beta P_{n-1} + R_1 R_2 C \frac{dQ_n}{dt}}{1 + \beta} \tag{2.17}$$

where $\beta = R_2 C / \Delta t$ is introduced for simplicity. Similarly, the flow derivative term can be approximated using the backward Euler method as follows

$$\frac{dQ_n}{dt} = \frac{Q_n - Q_{n-1}}{\Delta t} \tag{2.18}$$

which in combination with Eq. 2.15 yields

$$P_n = \frac{(R_1 + R_2 + R_1 \beta) Q_n + \beta P_{n-1} - R_1 \beta Q_{n-1}}{1 + \beta} \tag{2.19}$$

There are two approaches to implement these Equations in CFX: either by coupling with FORTRAN or by directly defining equations within the software-specific CEL language. The former technique utilises Eq. 2.19 and was used in Chap. 3. In the latter technique, the software is limited to storing one variable from the previous timestep, and can also provide the time derivative of a single variable. Thus Eq. 2.15 is used, with the time derivative of Q extracted from the software. This technique was developed for Chap. 5, in which a fluid-structure interaction approach was implemented, so as to avoid additional external couplings and thus increase efficiency.

It should be noted that if R_1 is set to 0, the system becomes a WK2 model and both Eqs. 2.17 and 2.19 reduce to

$$P_n = \frac{R_2 Q_n + \beta P_{n-1}}{1 + \beta} \tag{2.20}$$

which is the discretised version of Eq. 2.8.

2.4.4 Four-element Windkessel Models

In order to account for the inertial properties of the blood, the WK3 model can be further modified by adding an inductor either in parallel (WK4p) or in series (WK4s) with the characteristic impedance R_1. Kung and Taylor (2010) and Kung et al. (2011) showed that with a flexible tube and downstream 4-element Windkessel system, they

2.4 Dynamic Boundary Conditions

could simulate physiological pressures and flows. They subsequently replaced the tube with a reconstructed model of on abdominal aortic aneurysm and were still able to reproduce numerical and computational results (although they did not use blood, but rather a glycerol solution with dynamic viscosity of $\approx 4.6\,\text{mPa}\,\text{s}$) (Kung et al. 2011).

The addition of an extra parameter allows more freedom for the model to represent experimentally acquired data, and in theory to predict arterial waveforms. However, the additional parameters must be estimated. As the more complex circuits are not direct analogues, interpreting the physical meaning of a given parameter value for each component becomes poorly defined, given that the response of the circuit is dependent on all components. Furthermore, there will not be a unique combination of parameters for a given set of experimental data (Shi et al. 2011). Even for the WK2, the reported experimental and modelled values of R and C in the literature differ widely, and this is exacerbated with the additional complexity introduced by extra elements. Furthermore, if such models are to have clinical applications, simplicity and ease of use are paramount.

2.4.5 Compound Windkessel Models

Each of the models described above is capable of describing the lumped parameters of a single system. However, the vasculature can be separated into multiple sections, with each one represented by a WK model. For example, Westerhof et al. (1969) provided a complete 0D representation of the human arterial tree. In a more recent study, (Korakianitis and Shi 2006) used an RLC circuit for the aortic root in series with an RLC circuit for the systemic arteries, a single resistive element for the arterioles and capillaries, and an RC circuit to represent the venous side of the systemic vasculature. A similar arrangement was used for the pulmonary vasculature and the two were linked via a lumped parameter model for the heart, with active capacitors representing the atria and ventricles and diodes representing the heart valves. Figure 2.12 shows the left ventricular, venous and aortic pressures as a function of time, simulated this model using the CellML software (CellMl.org: accessed March 2013).

It can be seen that the system is able to produce physiologically realistic (although somewhat simplified) pressure waves at various points in the vasculature.

2.4.6 Parameters for Windkessel Models

In addition to their ability to model the vasculature using connected WK models, 0D models can also be utilised to provide BCs to 1, 2 or 3D models. When coupling domains of different dimensions, the problem of defective BCs (as a result of a reduction in the parameter space, e.g. cross sectional area) must be addressed

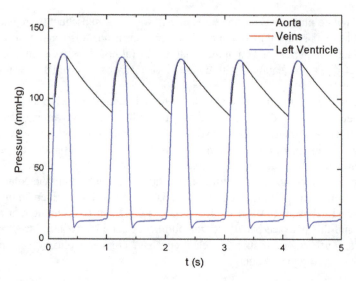

Fig. 2.12 Pressure traces at various locations simulated using 0D modelling (Korakianitis and Shi 2006)

appropriately, for example by assuming uniform pressure across the outlets of a 3D domain coupled to a 0D domain (Formaggia et al. 2008, 2010).

For use as BCs for 3D simulations, either 1D models (also known as impedance BCs) (Vignon-Clementel et al. 2006; Grinberg and Karniadakis 2008) or 0D models (Vignon-Clementel et al. 2006; Kim et al. 2009; Brown et al. 2012; Moghadam et al. 2013) can be used. The former is more accurate, and able to capture wave reflection characteristics, but require additional time to converge and can thus be very computationally expensive (Grinberg et al. 2009). The present work is concerned with the development of a clinically practicable tool, and thus focuses on 3D-0D coupling, which is more efficient and able to capture the essential characteristics of the downstream vasculature.

In this approach, Windkessel parameters need to be estimated for each domain outlet. There is no agreed methodology for parameter definition in RCR systems (Shi et al. 2011) with detailed descriptions of the methods used being surprisingly sparse. However, a large body of work has been devoted to fitting computational parameters to experimental data (Graham and Kilpatrick 2010; Moireau and Chapelle 2010; Perego et al. 2011). Bertoglio et al. (2011) used a Kalman filter approach to define peripheral resistance for an idealised abdominal aortic aneurysm.

A number of authors have used the 'coupled multi-domain method' (Vignon-Clementel et al. 2006; LaDisa et al. 2011). Total arterial compliance, estimated from measurements of the blood pressure and inflow for a given patient, is scaled empirically and defined at each outlet based on flow proportions. Resistances are then estimated based on proportions of the total peripheral resistance, to match the patient blood pressure. While the above method has the advantage of using clinical data to estimate compliance, it cannot be used for patients where flow data is not available.

2.4 Dynamic Boundary Conditions

Alternatively, iterative approaches, based on clinical measurements, can be used. Kim et al. (2009) coupled three-element windkessel models to the boundaries of a 3D aorta, in rest and exercise conditions, and for an aorta with a coarctation. Their Windkessel parameters were "adjusted to match subject-specific pulse pressure and cardiac output through an iterative approach". Similarly, in a study of computational efficiency in aortic simulations for the clinic, Brown et al. (2012) used three-element Windkessel models at the boundaries with parameters "tuned to match clinically measured pressure and flow data from the individual". However, no further details were provided in these papers.

In Chap. 3, a tuning methodology for estimating Windkessel parameters based on invasive pressure measurements is developed, in order to yield physiological BCs for the virtual treatment simulations and wall motion analyses in Chaps. 4 and 5 respectively. Details of the derivation of these parameters are given in Chap. 3. However, in the following section, the flow and pressure waves from Windkessel simulations are presented for the purpose of comparison, in order to demonstrate the importance of applying appropriate BCs.

2.4.7 Comparison of Zero-Pressure and Windkessel Boundary Conditions for Aortic Dissection

Given that one of the main themes of this thesis is that application of appropriate BCs is essential in simulating AD, it seems prudent to provide evidence to support this hypothesis. In this section, the model with Windkessel parameters is directly compared to results acquired if the outlet BCs are all set to zero. All other simulation parameters were the same as will be described in Chap. 3 (including the 3D geometry).

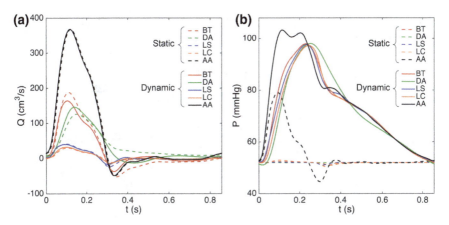

Fig. 2.13 Comparison of zero-pressure (*dashed lines*) and Windkessel boundary conditions (*solid lines*). **a** Flow waves and **b** pressure waves at the domain boundaries

Figure 2.13a compares the flow through each of the domain boundaries for Windkessel BCs (dynamic case) with solid lines and zero-pressure BCs (static case) with dashed lines.

The flow at the ascending aorta (AA), shown in black, is equal for both cases, as it is prescribed. The flow through the LS and LCC arteries is also very similar for the two simulations. However, the flow through the BT and DA differ considerably, particularly after $t \gtrsim 0.3$ s, wherein the backflow in the DA is not captured at all by the static simulation, and the BT has excessively high backflow. It can also be seen that the high frequency characteristics are not well captured in the static model. Given the simplicity of the static BC, it may be an appealing choice, as it approximates the flow to some extent. However, considering Fig. 2.13b, it is clear that the pressure distributions are not well captured by the simulation. A constant pressure of 52 mmHg, corresponding to the minimum pressure at the AA from the Windkessel parameters, is added to the static case to enable comparison. At the outlets, the pressure is approximately constant (slight deviations within the numerical resolution can be observed). However, pressure at the inlet is not prescribed, but calculated implicitly based on the simulation. It can be seen that the pressure pulse at the inlet in the static case fails entirely to model the true pressure pulse; rather, it represents the pressure gradient between the inlet and the outlets. To illustrate this more clearly, Fig. 2.14 plots the pressure at each of the outlet branches relative to that at the inlet. It can be seen that for the static model, the lines collapse onto a single curve, equivalent to the inlet pressure in the static case (Fig. 2.13b). For the dynamic simulation, there is a range of pressure differences around peak systole: a result of the phase differences between the different branches, introduced by the different impedances in the Windkessel models. Particularly, the pressure gradient

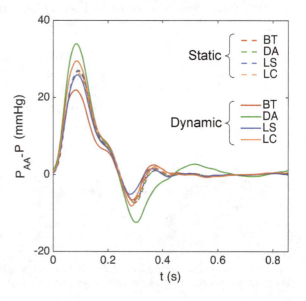

Fig. 2.14 Pressure for each of the domain outlets relative to the pressure at the AA for zero-pressure (*dashed lines*) and Windkessel (*solid lines*) boundary conditions

at the distal abdominal for the dynamic case (and thus in the descending aorta, the region of the aorta which is of most interest for type-B AD), deviates significantly from the static case.

Thus, it can be concluded that if one is interested in the flow in the supraaortic branches only, or in purely time averaged parameters (such as averaged shear indices), then zero-pressure BCs might be sufficient. However, where pressure distributions, streamlines, oscillatory shear indices etc. are of interest, zero pressure BCs are unable to fully capture the true dynamics of the system. Given that the utility of these simulations as clinical tools relies on their ability to produce physiologically realistic predictions of the haemodynamics, it is clear that dynamic BCs are required.

2.4.8 Comparison of Flow-Split and Windkessel Boundary Conditions for Aortic Dissection

The simplest approach to modelling dynamic BCs, is to assume that the flow through the aortic branches is a given proportion of the inlet flow, generally 5% (Chen et al. 2013a, b; Tse et al. 2011). For example, for the case of the BT, this can be stated mathematically as $\overline{Q_{BT}} = 0.05\overline{Q_{AA}}$. However, in the transient simulations of AD, this is prescribed as $Q_{BT}(t) = 0.05 Q_{AA}(t)$. The implications of this are not commonly mentioned, but in translating the proportion of averaged flow to the individual time-steps, the phases of all of the branches are forced to be equal. Figure 2.15a compares the flow calculated with the Windkessel and flow-split cases. In the latter, each of the supraaortic branch flow rates was scaled to 5% of the inlet flow. The outlet flow was thus implicitly set to be 85% of the inlet flow. In order to provide a reference pressure for the simulation, the outlet pressure was set to 0 mmHg, following Chen et al. (2013a). Comparing the Windkessel and flow-split results for the DA, it is clear that they are very different, and the flow-split condition results in almost twice the amount of flow compared to the Windkessel simulation. Correspondingly, the flows in the BT, LS and LC (the latter covers the others in Fig. 2.15a as they are prescribed to be identical) are very small and all equal, as compared to the Windkessel simulation. Without clinical flow data, it cannot be explicitly demonstrated which model is superior. However, as can be seen in Fig. 2.15b, the pressure waves required to produce such a flow are non-physiological. Not only are the pressure waves at each boundary completely in phase, but the pulse pressure is more than doubled. This is due to the fact that a very high flow rate must be forced through the coarctation, and the TL, which is greatly reduced in diameter as compared to a healthy aorta.

Figure 2.16 highlights this issue, showing a peak pressure difference between the inlet and DA at peak systole of >90 mmHg (for the flow-split case). Conversely, in the branches, the peak pressure drop is around 10 mmHg as compared to 20–30 mmHg for the Windkessel model.

Thus, flow-split is not an appropriate choice of BCs for modelling AD, as the predicted results are not realistic, and are thus of limited clinical use.

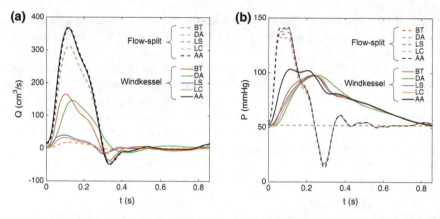

Fig. 2.15 Comparison of flow-split (*dashed lines*) and Windkessel boundary conditions (*solid lines*). **a** Flow waves and **b** pressure waves at the domain boundaries

Fig. 2.16 Pressure for each of the domain outlets relative to the pressure at the AA for flow-split (*dashed lines*) and Windkessel (*solid lines*) boundary conditions

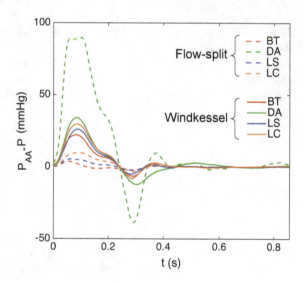

Figures 2.13, 2.14, 2.15 and 2.16 provide incontrovertible evidence that BCs for 3D simulations of AD must not only be dynamic, but also capable of accounting for phase differences in the pressures/flows through each branch.

2.5 Finite Element Modelling

Finite element modelling is used in the present study in the context of fluid-structure interaction in order to model the vessel wall motion in AD and evaluate its influence on haemodynamics. In this section, a brief introduction to the topic will be given. Further details will be provided in Chap. 5.

2.5.1 Vessel Wall Reconstruction

The first step in carrying out a finite element analysis is to generate the geometry. The limited spatial resolution of clinical image modalities limits the ability to resolve the vessel wall thickness directly from imaging data (Speelman et al. 2008; Erbel et al. 2001). The simplest approach to generate the wall is to extrude the outer wall of the fluid domain by a fixed amount (Speelman et al. 2008; Qiao et al. 2014; Nathan et al. 2011). The thickness of the vessel wall, however, varies throughout the domain. One approach is to independently define the thickness of the supraaortic branches to be 1 mm, with 2 mm elsewhere (Moireau et al. 2011, 2013). An alternative methodology is to define the wall thickness as a constant ratio of the lumen diameter (Crosetto et al. 2011; Reymond et al. 2013), i.e. $h = 0.071 D_{\text{lumen}}$.

2.5.2 Material Properties

The simplest material model for simulating solid mechanics is a linear elastic material, in which the stress, σ, and strain, ϵ, are proportional to one another,

$$\sigma = E\epsilon \qquad (2.21)$$

where E is the Young's modulus. FSI studies of the aorta using linear elastic models have used a range of Young's moduli, such as 0.4 MPa (Crosetto et al. 2011; Reymond et al. 2013; Colciago et al. 2014), 0.63 MPa (Kim et al. 2009), 0.7 MPa (Xiao et al. 2013), 0.84 MPa (Gao et al. 2006), 1 MPa (Brown et al. 2012) and 3 MPa (Nathan et al. 2011).

Raghavan and Vorp (2000) analysed the material properties of a large number of abdominal aortic aneurysms, and reported deviation from linear elastic behaviour. They also observed that longitudinal and circumferential sections of tissue had similar parameters, and thus the assumption of isotropy was appropriate. They derived a model from principles of finite strain theory, which are beyond the scope of the present work. However, with the isotropic assumption, the material behaviour can be considered in the context of uniaxial loading, in which the stress and stretch, $\lambda = \epsilon + 1$, are related according to

$$\sigma = \left(2A + 4B\left(\lambda^2 + 2/\lambda - 3\right)\right)\left(\lambda^2 - 1/\lambda\right) \qquad (2.22)$$

where A and B are empirically defined constants. Figure 2.17 compares the stress-strain behaviour of the model described by Eq. 2.22, as compared to a number of linear elastic models.

For the simulations of healthy aortae (Brown et al. 2012; Reymond et al. 2013), the relationship is linear, although the slopes are rather different, indicating a much stiffer material used by Brown et al. (2012). The study of Nathan et al. (2011) also used a linear elastic model, but to represent the increased stiffness observed in aortic

Fig. 2.17 Comparison of uniaxial stress-strain behaviour with different material models

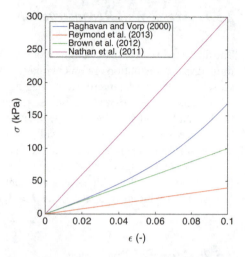

disease, a higher Young's modulus was employed. It can be seen that the model of Raghavan and Vorp (2000) is non-linear, and for higher strains, an increasing amount of stress is required to induce further deformations: the material exhibits load-stiffening and thus becomes stiffer as it deforms. However, for small strains (<0.02) the model of Raghavan and Vorp (2000) is very similar to the linear elastic model used by Brown et al. (2012).

An additional property of the walls of arteries is that they exist in a state of pre-stress, (Chuong and Fung 1986; Fung 1991). The effect of pre-stress on Fig. 2.17 would be to shift the blue curve to the left, enhancing the load-stiffening behaviour at low strains.

Although the linear elastic approach is a considerable simplification, neglecting the influence of the multi-layered structure, anisotropy and non-linearity of the vessel wall, it is able to produce meaningful wall displacements (Crosetto et al. 2011), although it cannot accurately represent internal stresses (Reymond et al. 2013). The non-linear behaviour captured by Eq. 2.22 more accurately predicts stress distributions than a linear elastic model (Raghavan and Vorp 2000), but underestimates peak stresses relative to anisotropic models (Roy et al. 2014).

There is an increasing body of work on modelling the vessel wall as anisotropic, considering the alignment of collagen fibers and multiple layers of the arterial wall, as reviewed by Gasser et al. (2006). The accuracy of such models is increased compared to isotropic simplifications, but they require considerable additional computational effort as well as experimental data for validation, which is not presently available in the context of AD. In a patient-specific context, the absence of direct measurements of wall properties adds additional complexity to modelling the vessel wall, even with isotropic linear-elastic models. In Chap. 5, an FSI simulation of AD will be described.

The following chapter proposes a methodology for iteratively deriving Windkessel parameters, based on invasive pressure measurements. These BCs are then applied in Chaps. 4 and 5 to investigate treatment options and the influence of wall motion respectively.

References

ANSYS (2011), *CFX Solver Theory Guide*, 14.0 edition.

Patankar, S. V. (1980). *Numerical heat transfer and fluid flow*. New York: McGraw-Hill.

Barth, T., & Jespersen, D. (2012). The design and application of upwind schemes on unstructured meshes, in *27th Aerospace Sciences Meeting*. Reston, Virigina: American Institute of Aeronautics and Astronautics.

Bertoglio, C., Moireau, P., & Gerbeau, J.-F. (2011). Sequential parameter estimation for fluid-structure problems: Application to hemodynamics. *International Journal for Numerical Methods in Biomedical Engineering*, 28(4), 434–455.

Beynon, R., Sterne, J. A. C., Wilcock, G., Likeman, M., Harbord, R. M., Astin, M., et al. (2012). Is MRI better than CT for detecting a vascular component to dementia? A systematic review and meta-analysis. *BMC Neurology*, 12(1), 33.

Brown, A. G., Shi, Y., Marzo, A., Staicu, C., Valverde, I., Beerbaum, P., et al. (2012). Accuracy vs. computational time translating aortic simulations to the clinic. *Journal of Biomechanics*, 45(3), 516–523.

Cheng, Z., Juli, C., Wood, N. B., Gibbs, R. G. J., & Xu, X. Y. (2014). Predicting flow in aortic dissection: Comparison of computational model with PC-MRI velocity measurements. *Medical Engineering and Physics*, 36(9), 1176–1184.

Chen, D., ller Eschner, M. M., von Tengg-Kobligk, H., Barber, D., Bockler, D., Hose, R., et al. (2013a). A patient-specific study of type-B aortic dissection: evaluation of true-false lumen blood exchange. *BioMedical Engineering OnLine*, 12, 65.

Chen, D., Müller-Eschner, M., Kotelis, D., Böckler, D., Ventikos, Y., & von Tengg-Kobligk, H. (2013b). A longitudinal study of Type-B aortic dissection and endovascular repair scenarios: Computational analyses. *Medical Engineering and Physics*, 35(9), 1321–1330.

Chuong, C. J., & Fung, Y. C. (1986). On residual stresses in arteries. *Journal of Biomechanical Engineering*, 108(2), 189–192.

JCS Joint Working Group (2013). Guidelines for diagnosis and treatment of aortic aneurysm and aortic dissection (JCS 2011). *Circulation Journal*, 77(3), 789–828.

Clough, R. E., Waltham, M., Giese, D., Taylor, P. R., & Schaeffter, T. (2012). A new imaging method for assessment of aortic dissection using four-dimensional phase contrast magnetic resonance imaging. *Journal of Vacscular Surgery*, 55(4), 914–923.

Colciago, C. M., Deparis, S., & Quarteroni, A. (2014). Comparisons between reduced order models and full 3D models for fluid-structure interaction problems in haemodynamics. *Journal of Computational and Applied Mathematics*, 265, 120–138.

Crosetto, P., Reymond, P., Deparis, S., & Kontaxakis, D. (2011). Fluid-structure interaction simulation of aortic blood flow. *Computers and Fluids*, 43, 46–57.

Decorato, I., Salsac, A.-V., Legallais, C., Alimohammadi, M., Díaz-Zuccarini, V., & Kharboutly, Z. (2014). Influence of an arterial stenosis on the hemodynamics within an arteriovenous fistula (AVF): Comparison before and after balloon-angioplasty. *Cardiovascular Engineering and Technology*, 5(3), 233–243.

Erbel, R., Alfonso, F., Boileau, C., Dirsch, O., Eber, B., Haverich, A., et al. (2001). Diagnosis and management of aortic dissection task force on aortic dissection, european society of cardiology. *European Heart Journal*, 22(18), 1642–1681.

Formaggia, L., Veneziani, A., & Vergara, C. (2008). A new approach to numerical solution of defective boundary value problems in incompressible fluid dynamics. *SIAM Journal on Numerical Analysis*, 46(6), 2769–2794.

Formaggia, L., Veneziani, A., & Vergara, C. (2010). Comput. Methods Appl. Mech. Engrg. *Computer Methods in Applied Mechanics and Engineering*, 199(9–12), 677–688.

Francois, C. J., Markl, M., Schiebler, M. L., Niespodzany, E., Landgraf, B. R., Schlensak, C., et al. (2013). Four-dimensional, flow-sensitive magnetic resonance imaging of blood flow patterns in thoracic aortic dissections. *The Journal of Thoracic and Cardiovascular Surgery*, 145(5), 1359–1366.

Fung, Y. C. (1991). What are the residual stresses doing in our blood vessels? *Annals of Biomedical Engineering, 19*(3), 237–249.

Fung, Y. (1993). *Biomechanics: Mechanical properties of living tissues* (2nd ed.). Heidelberg: Springer.

Fung, Y. (1997). *Biomechanics: Circulation* (2nd ed.). New York: Springer.

Ganten, M.-K., Weber, T. F., von Tengg-Kobligk, H., Böckler, D., Stiller, W., Geisbüsch, P., et al. (2009). Motion characterization of aortic wall and intimal flap by ECG-gated CT in patients with chronic B-dissection. *European Journal of Radiology, 72*(1), 146–153.

Gao, F., Guo, Z., Sakamoto, M., & Matsuzawa, T. (2006). Fluid-structure Interaction within a layered aortic arch model. *Journal of Biological Physics, 32*(5), 435–454.

Gasser, T. C., Ogden, R. W., & Holzapfel, G. A. (2006). Hyperelastic modelling of arterial layers with distributed collagen fibre orientations. *Journal of The Royal Society Interface, 3*(6), 15–35.

Graham, L. S., & Kilpatrick, D. (2010). Estimation of the bidomain conductivity parameters of cardiac tissue from extracellular potential distributions initiated by point stimulation. *Annals of Biomedical Engineering, 38*(12), 3630–3648.

Grinberg, L., Anor, T., Madsen, J. R., Yakhot, A., & Karniadakis, G. E. (2009). Large-scale simulation of the human arterial tree. *Clinical and Experimental Pharmacology and Physiology, 36*(2), 194–205.

Grinberg, L., & Karniadakis, G. E. (2008). Outflow boundary conditions for arterial networks with multiple outlets. *Annals of Biomedical Engineering, 36*(9), 1496–1514.

Karmonik, C., Duran, C., Shah, D. J., Anaya-Ayala, J. E., Davies, M. G., Lumsden, A. B., et al. (2012a). Preliminary findings in quantification of changes in septal motion during follow-up of type B aortic dissections. *Journal of Vacscular Surgery, 55*(5), 1419.e1–1426.e1.

Karmonik, C., Partovi, S., Davies, M. G., Bismuth, J., Shah, D. J., Bilecen, D., et al. (2012b). Integration of the computational fluid dynamics technique with MRI in aortic dissections. *Magnetic Resonance in Medicine, 69*(5), 1438–1442.

Kim, H. J., Vignon-Clementel, I. E., Figueroa, C. A., LaDisa, J. F., Jansen, K. E., Feinstein, J. A., et al. (2009). On coupling a lumped parameter heart model and a three-dimensional finite element aorta model. *Annals of Biomedical Engineering, 37*(11), 2153–2169.

Korakianitis, T., & Shi, Y. (2006). Numerical simulation of cardiovascular dynamics with healthy and diseased heart valves. *Journal of Biomechanics, 39*(11), 1964–1982.

Kung, E. O., Les, A. S., Figueroa, C. A., Medina, F., Arcaute, K., Wicker, R. B., et al. (2011a). In vitro validation of finite element analysis of blood flow in deformable models. *Annals of Biomedical Engineering, 39*(7), 1947–1960.

Kung, E. O., Les, A. S., Medina, F., Wicker, R. B., McConnell, M. V., & Taylor, C. A. (2011b). In vitro validation of finite-element model of AAA hemodynamics incorporating realistic outlet boundary conditions. *Journal of Biomechanical Engineering, 133*(4), 041003.

Kung, E. O., & Taylor, C. A. (2010). Development of a physical windkessel module to re-create in vivo vascular flow impedance for in vitro experiments. *Cardiovascular Engineering and Technology, 2*(1), 2–14.

LaDisa, J. F, Jr., Dholakia, R. J., Figueroa, C. A., Vignon-Clementel, I. E., Chan, F. P., Samyn, M. M., et al. (2011). Computational simulations demonstrate altered wall shear stress in aortic coarctation patients treated by resection with end-to-end anastomosis. *Congenital Heart Disease, 6*(5), 432–443.

Layton, K. F., Kallmes, D. F., Cloft, H. J., Lindell, E. P., & Cox, V. S. (2006). Bovine aortic arch variant in humans: Clarification of a common misnomer. *American Journal of Neuroradiology, 27*(7), 1541–1542.

Levick, J. R. (2009). *An indtroduction to cardiovascular physiology* (5th ed.). London: Hodder Arnold.

Moghadam, M. E., Vignon-Clementel, I. E., Figliola, R., & Marsden, A. L. (2013). A modular numerical method for implicit 0D/3D coupling in cardiovascular finite element simulations. *Journal of Computational Physics, 244*(C), 63–79.

References

Moireau, P., Bertoglio, C., Xiao, N., Figueroa, C. A., Taylor, C. A., Chapelle, D., et al. (2013). Sequential identification of boundary support parameters in a fluid-structure vascular model using patient image data. *Biomechanics and Modeling in Mechanobiology, 12*(3), 475–496.

Moireau, P., & Chapelle, D. (2010). Reduced-order Unscented Kalman Filtering with application to parameter identification in large-dimensional systems. *ESAIM: Control Optimisation and Calculus of Variations, 17*(2), 380–405.

Moireau, P., Xiao, N., Astorino, M., Figueroa, C. A., Chapelle, D., Taylor, C. A., et al. (2011). External tissue support and fluid-structure simulation in blood flows. *Biomechanics and Modeling in Mechanobiology, 11*, 1–18.

Munson, B., Young, D., & Okiishi, T. (1994). *Fundamentals of fluid mechanics* (2nd ed.). New York: Wiley.

Nathan, D. P., Xu, C., Gorman, J. H, I. I. I., Fairman, R. M., Bavaria, J. E., Gorman, R. C., et al. (2011). Pathogenesis of acute aortic dissection: A finite element stress analysis. *The Annals of Thoracic Surgery, 91*(2), 458–463.

Perego, M., Veneziani, A., & Vergara, C. (2011). A variational approach for estimating the compliance of the cardiovascular tissue: An inverse fluid-structure interaction problem. *SIAM Journal on Scientific Computing, 33*(3), 1181–1211.

Peterson, K. L., Tsuji, J., Johnson, A., DiDonna, J., & LeWinter, M. (1978). Diastolic left ventricular pressure-volume and stress-strain relations in patients with valvular aortic stenosis and left ventricular hypertrophy. *Circulation, 58*(1), 77–89.

Qiao, A., Yin, W., & Chu, B. (2014). Numerical simulation of fluid-structure interaction in bypassed DeBakey III aortic dissection. *Computer Methods in Biomechanics and Biomedical Engineering, 18*(11), 1173–1180.

Raghavan, M. L., & Vorp, D. A. (2000). Toward a biomechanical tool to evaluate rupture potential of abdominal aortic aneurysm: Identification of a finite strain constitutive model and evaluation of its applicability. *Journal of Biomechanics, 33*, 475–482.

Reymond, P., Crosetto, P., Deparis, S., Quarteroni, A., & Stergiopulos, N. (2013). Physiological simulation of blood flow in the aorta: Comparison of hemodynamic indices as predicted by 3-D FSI, 3-D rigid wall and 1-D models. *Medical Engineering and Physics, 35*(6), 784–791.

Rhie, C. & Chow, W. (2012). A numerical study of the turbulent flow past an isolated airfoil with trailing edge separation. In *3rd Joint Thermophysics, Fluids, Plasma and Heat Transfer Conference*. Reston, Viriginia: American Institute of Aeronautics and Astronautics.

Roy, D., Holzapfel, G. A., Kauffmann, C., & Soulez, G. (2014). Finite element analysis of abdominal aortic aneurysms: Geometrical and structural reconstruction with application of an anisotropic material model. *IMA Journal of Applied Mathematics, 79*(5), 1011–1026.

Shi, Y., Lawford, P., & Hose, R. (2011). Review of Zero-D and 1-D models of blood flow in the cardiovascular system. *BioMedical Engineering OnLine, 10*(1), 33.

Speelman, L., Bosboom, E. M. H., Schurink, G. W. H., Hellenthal, F. A. M. V. I., Buth, J., Breeuwer, M., et al. (2008). Patient-specific AAA wall stress analysis: 99-percentile versus peak stress. *European Journal of Vascular and Endovascular Surgery, 36*(6), 668–676.

Stergiopulos, N., Westerhof, B. E., & Westerhof, N. (1999). Total arterial inertance as the fourth element of the windkessel model. *American Journal of Physiology-Legacy Content, 276*(1), H81–H88.

Studwell, A. J., & Kotton, D. N. (2011). A shift from cell cultures to creatures: In vivo imaging of small animals in experimental regenerative medicine. *Molecular Therapy, 19*(11), 1933–1941.

Tan, F. P. P., Borghi, A., Mohiaddin, R. H., Wood, N. B., Thom, S., & Xu, X. Y. (2009). Analysis of flow patterns in a patient-specific thoracic aortic aneurysm model. *Computers and Structures, 87*(11–12), 680–690.

Taylor, A., Sheridan, M., McGee, S., & Halligan, S. (2005). Preoperative staging of rectal cancer by MRI; results of a UK survey. *Clinical Radiology, 60*, 579–586.

Tse, K. M., Chiu, P., Lee, H. P., & Ho, P. (2011). Investigation of hemodynamics in the development of dissecting aneurysm within patient-specific dissecting aneurismal aortas using computational fluid dynamics (CFD) simulations. *Journal of Biomechanics, 44*(5), 827–836.

Versteeg, H., & Malalasekera, W. (2007). *An introduction to computational fluid dynamics* (2nd ed.). Upper Saddle River: Prentice Hall.

Vignon-Clementel, I. E., Alberto Figueroa, C., Jansen, K. E., & Taylor, C. A. (2006). Outflow boundary conditions for three-dimensional finite element modeling of blood flow and pressure in arteries. *Computer Methods in Applied Mechanics and Engineering, 195*(29–32), 3776–3796.

Wesseling, P. (2000). *Principles of compuational fluid dynamics*. Heidelberg: Springer.

Westerhof, N., Bosman, F., De Vries, C. J., & Noordergraaf, A. (1969). Analog studies of the human systemic arterial tree. *Journal of Biomechanics, 2*(2), 121–143.

Westerhof, N., Lankhaar, J.-W., & Westerhof, B. E. (2008). The arterial Windkessel. *Medical and Biological Engineering and Computing, 47*(2), 131–141.

White, F. (2011). *Fluid mechanics* (7th ed.). New York: McGraw-Hill.

Xiao, N., Alastruey, J., & Alberto Figueroa, C. (2013). A systematic comparison between 1-D and 3-D hemodynamics in compliant arterial models. *International Journal for Numerical Methods in Biomedical Engineering, 30*(2), 204–231.

Yang, S., Li, X., Chao, B., Wu, L., Cheng, Z., Duan, Y., et al. (2014). Abdominal aortic intimal flap motion characterization in acute aortic dissection: Assessed with retrospective ECG-Gated thoracoabdominal aorta dual-source CT angiography. *PLoS ONE, 9*(2), e87664.

Chapter 3
Haemodynamics of a Dissected Aorta

In this chapter, a three-element Windkessel model is coupled to each of the outflow boundaries of a 3D geometry from a patient with a type-B AD. A novel methodology is developed to define the parameters of the dynamic boundary conditions for the simulation. The Windkessel parameters are tuned iteratively to match maxima and minima of invasive pressure measurements taken from the same patient. The results of the simulation indicate blood flow reduction for the downstream organs and highlight interesting haemodynamics in the dissected region.

3.1 Introduction

The contraction of the heart ejects blood periodically, resulting in flow and pressure waves that propagate through the vasculature and reflect as a result of downstream tapering, bifurcations etc. The vessels bifurcate sequentially, resulting in thousands of progressively smaller diameter vessels. When simulating the large vessels of the vasculature, it is necessary to reduce the model domain, since accounting for all of the downstream and upstream vessels is not only computationally intractable but also difficult to define. In this chapter, only a limited section of the vasculature (the region

---

The work presented in this chapter was published in 'Development of a patient-specific simulation tool to analyse aortic dissections: Assessment of mixed patient-specific flow and pressure boundary conditions', *Medical Engineering and Physics* (Alimohammadi et al. 2014a). This work was also presented at the ESB Biomechanics congress, Lisbon, 2012 and the IMA Mathematics of Medical Devices and Surgical Procedures conference, London, 2012. The techniques developed in this chapter were also used in a further study: 'Predicting atherosclerotic plaque location in an iliac bifurcation using a hybrid CFD/biomechanical approach', *Lecture Notes in Computer Science* (Alimohammadi et al. 2015).

of interest) will be modelled in 3D and will be coupled to three-element Windkessel models of the downstream vasculature (Fig. 2.7).

As discussed in Chaps. 1 and 2, one of the most important factors in patient-specific simulations is to prescribe appropriate BCs. In the real system, pressure and flow at each domain boundary vary throughout the cardiac cycle. Considering the time dependence of BCs is particularly important in the case of AD, as the vessel narrowing in the coarctation (narrowed region in the distal aortic arch) and along the TL, in addition to the obstruction of downstream vessels due to the motion of the intimal flap, affect the flow direction and proportion through each branch (static and dynamic obstruction) (Swee and Dake 2008). Malperfusion of the downstream vessels and organs is one of the most important decision-making parameters that clinicians consider when monitoring AD (Erbel et al. 2001); however, currently these parameters can only be verified by laboratory markers or imaging (Fattori et al. 2013).

In recent years, researchers have conducted a number of studies to simulate blood flow through a patient-specific ADs. In the absence of patient-specific data, constant or zero pressure has commonly been deployed at the outlets of the patient-specific 3D domain (Cheng et al. 2010, 2013; Karmonik et al. 2011b) or a specific flow proportion prescribed for each outlet (Tse et al. 2011). Chen et al. (2013b) used time-varying flow and pressure distributions that were extracted from other studies. Although it is necessary to make certain assumptions in the absence of complete information at the boundaries, these simplifications are clearly detrimental to the accuracy, reliability and applicability of the simulations, as shown in Chap. 2.

Where non-invasive data such as flow or pressure waves are available, they can be directly applied at the appropriate boundary. A number of AD studies have applied an inflow waveform, measured with 2D pcMRI at the inlet of the 3D domain (Karmonik et al. 2008, 2012a). However, downstream obstruction and coarctation regions in AD alter and disturb the flow behaviour, and the proportion of flow entering each branch may vary from the values found in healthy aortae. Therefore, special care is required in order to select appropriate time varying flow or pressure waves at the outlets of the 3D domain. Coupling CFD simulations with Windkessel or zero-dimensional (0D) models offers an alternative methodology that accounts for the interdependent time-varying flow and pressure distribution, and hence is more suitable in patient specific simulations (Blanco et al. 2007; Brown et al. 2012; Formaggia et al. 1999; Kim et al. 2009; Quarteroni et al. 2002; Suh et al. 2010). This approach has not been reported previously in the context of AD and will be the main focus of this chapter. The superiority of dynamic BCs over zero-pressure or flow split conditions was demonstrated in the previous chapter.

3.2 Methodology

Details of the meshing procedure were provided in Chap. 2. Briefly, the extracted AD geometry was meshed using ANSYS ICEM-CFD (14.0). The domain was discretised into approximately 250,000 tetrahedral elements with 7 prism layers at the walls.

3.2 Methodology

Sample images of the mesh can be found in Appendix A. In addition, a coarse mesh with ≈80,000 elements and a fine mesh with ≈600,000 elements, with 4 and 7 prism layers respectively, were created to assess the mesh sensitivity.

At each of the domain outlets (BT, LS, LCC, DA) Windkessel models were implemented to represent the downstream vasculature. The inlet flow wave was not available for this patient, so flow from a dissected aorta in a previous study was utilised (Karmonik et al. 2008) and interpolated to the heart rate of the patient (70 beats per minute). The vessel wall was assumed to be rigid, and a no-slip condition was applied at the wall. At each of the domain boundaries, Windkessel models were utilised. The Windkessel models were implemented via FORTRAN sub-routines in ANSYS-CFX (ANSYS Inc., Canonsburg, PA, USA). For each internal solver loop, CFX passed the instantaneous flow rate to FORTRAN, calculated according to:

$$Q_i = \int_{A_i} \vec{u}.\vec{n}_i \, dA \quad (3.1)$$

where $\vec{u}$ is the velocity vector, $\vec{n}_i$ is the normal vector for interface i (constant for each interface) and A_i is the area of the interface. Hence, $Q_{3D,i} = Q_{0D,i}$, as required for the interface to be appropriate (Formaggia et al. 2009). The instantaneous pressure, $P_{0D,i}(t)$ was calculated based on Eq. 2.19, using the pressure and flow from the previous time step, $P_{0D,i}(t - \Delta t)$ and $Q_{3D,i}(t - \Delta t)$ respectively, and the flow rate for the given solver loop, $Q_{3D,i}(t)$. The calculated pressure was passed back to CFX, and was applied as a uniform BC in the subsequent solver loop.

3.2.1 Boundary Conditions and Data Assimilation Method

In order to define realistic values for each parameter in the three-element Windkessel models (R_{1i}, R_{2i} and C_i), invasive pressure measurements were carried out: a standard clinical routine once an intervention is determined to be necessary. Informed consent and approval from the ethical committee were obtained by the managing clinician. All measurements were obtained following standard clinical procedures undertaken by the clinical team. A transfemoral 5 French sized universal flush angiographic catheter (Cordis Corporation, Bridgewater, NJ, USA) with a radio-opaque tip and multiple distal openings, connected to a pressure transducer, was used for the measurements. Pressure readings were captured on the anaesthetic monitor in the hybrid endovascular theatre at University College London Hospital (UCLH). Measurements were acquired in the distal aortic arch, mid abdominal aorta and at each of the domain boundaries, with the exception of the LS branch, as it was deemed unsafe by the clinician to take measurements so close to the aneurysm. For the same reason, pressure measurements were not acquired in the FL. It should be noted that the pressure distribution inside the FL is unknown and cannot be measured with the current advanced imaging methods (Francois et al. 2013). For each measurement location, a minimum of 5 cardiac cycles were averaged and the minimum and

Table 3.1 Minimum and maximum pressure values at various locations throughout the domain, measured clinically and derived from the simulations (discussed later in this section). All pressures in mmHg. Difference rows show the difference between the measured and simulated values as a percentage of the AA pulse pressure

		AA	BT	LCC	LS	DA	DAA	MA
Clinical (mmHg)	P_{min}	56	50	50	N/A	52	53	52
	P_{max}	103	97	96	N/A	99	100	100
CFD (mmHg)	P_{min}	52.7	52	51.9	52.3	51.1	52.5	52.3
	P_{max}	103.3	97.9	97.3	97.6	98.1	101.2	101.3
Difference(%)	P_{min}	-7.0	4.3	4.0	N/A	-1.9	-1.1	0.6
	P_{max}	0.6	1.9	2.8	N/A	-1.9	2.6	2.8

maximum pressures were recorded. It was not possible to record the complete pressure waves for the present patient, and hence only the minima and maxima of the pressure at each location (the domain flow boundaries) were available for this investigation. Table 3.1 summarises the acquired data.

It can be observed that the measured values are relatively low, which is probably due to the patient's state of anaesthesia. However, this is part of the standard clinical procedure and thus cannot be avoided. Furthermore, it is the pressure drop, as opposed to the absolute pressure, that defines the flow through the vessel in the absence of interaction between the fluid and the mechanical properties of the vessel wall; thus the relatively lower magnitude of the measured pressures are not expected to affect the predicted haemodynamic parameters under the present simulation conditions.

In order to define the RCR parameters of the three-element Windkessel models, a data assimilation technique was applied. The aim was to minimise the mismatch, M, between the simulations and the invasively measured data:

$$\min_{R_{1i}, R_{2i}, C_i} M_i \quad (3.2)$$

$$M_i = \sqrt{\left(P_{min,clin} - P_{min,comp}\right)^2 + \left(P_{max,clin} - P_{max,comp}\right)^2} \quad (3.3)$$

Where R_{1i}, R_{2i} and C_i are the parameters and M_i is the mismatch (Eq. 3.3) at each of the interfaces, i. A number of rigorous mathematical approaches for minimising discrepancies between experimental and computational data have been reported in the literature, based on a variational approach (D'Elia et al. 2011) or a sequential approach utilising a Kalman filter (Bertoglio et al. 2011). In such approaches, the system is solved monolithically with the 3D velocity and pressure fields and parameters of interest optimised simultaneously. However, in the absence of full time resolved pressure measurements, continuous optimisation cannot be used. In the present case, a 'splitting' approach is utilised. Using this method, initial 3D simulations with zero pressure at the boundaries are performed in order to generate flow data, from which RCR parameters are calculated for each outlet of the 3D domain according to

3.2 Methodology

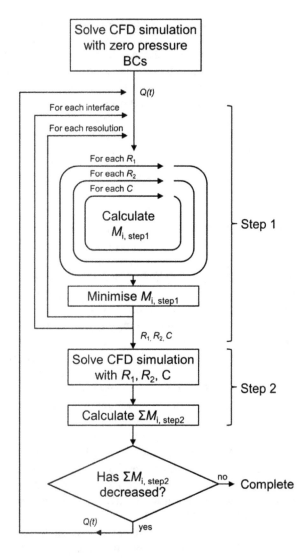

Fig. 3.1 Block diagram showing the approach used to tune the Windkessel parameters

Eq. 2.15 using LabVIEW (National Instruments, Austin, TX, USA). The calculated parameters are then used in a subsequent 3D simulation and the pressures at the boundaries extracted. The mismatch, M, between the clinical and numerical data is then evaluated and the process is repeated until M is not reduced by further iterations.

This iterative, two-step approach employed in order to obtain appropriate Windkessel parameters is shown diagrammatically in Fig. 3.1. $M_{i,\text{step1}}$ is defined according to Eq. 3.3, with P_{comp} defined by numerically solving the three element Windkessel equation (Eq. 2.15) in LabVIEW. $M_{i,\text{step2}}$ is defined according to Eq. 3.3 with P_{comp} extracted from the CFD simulations.

As only P_{max} and P_{min} were available, there is no unique combination of RCR parameters that will reproduce these values for a given flow wave. Hence, the following approach was utilised in order to improve the repeatability of the algorithm.

To provide an original estimate of the flow at each outlet, a simulation with $0\,Pa$ boundary conditions was run. Due to the transient response of RCR circuits, the flow waveform at each outlet was repeated so as to create an input profile of 50 cardiac cycles. Based on the steady-state solution of an RCR circuit and the assumption that $R_1 \ll R_2$ (Shi et al. 2011), an initial estimate for R_2 can be made according to $R_{2i}^0 = \overline{P}/\overline{Q}$, where the overbar represents the average over a cardiac cycle. $\overline{P}$ was estimated based on the standard formula (Levick 2009), $\overline{P} = P_{min} + (P_{max} - P_{min})/3$. $\overline{Q}$ was calculated by integrating the flow waveforms extracted from the $0\,Pa$ BC simulations. The range of R_2 was then considered to be $R_{2i}^0 \pm 1\,\text{mmHg s/ml}$. Initial ranges of $0 < R_1 < 1$ mmHg s/ml and $0 < C < 2$ ml/mmHg were defined (Brown et al. 2012; Kim et al. 2009). Initial search resolutions of 0.1 mmHg s/ml, for the resistances and 0.1 ml/mmHg for the compliance were utilised. The minimisation algorithm was designed so that for each combination of parameters, the pressure based on Eq. 2.9 was calculated, from which P_{max} and P_{min} were recorded and used to calculate the mismatch, M (Eq. 3.3). The parameter combination with the smallest M was selected. The resolution was then multiplied by 0.1. The search range for each parameter was limited to the best fit plus or minus the previous resolution. This process was carried out for each of the four outlet branches. In the LS, wherein invasive pressure data were unavailable, the pressure was assumed to be equivalent to that in the LCC for tuning purposes. For all 4 Windkessel interfaces, this process took less than 5 min.

The calculated RCR parameters were subsequently utilised in the 3D simulation. The pressure at each outlet was initialised based on the pressure at the beginning of the cardiac cycle calculated in the mismatch minimisation stage. This obviated the need to wait for a large number of cardiac cycles in order to reach a periodic response, due to the effect of the capacitance elements in the Windkessel circuits. Sufficient periodicity was achieved (maximum and minimum pressures changed by less than 1% over a cardiac cycle) within two cardiac cycles, after which the simulation was stopped, the flow at each boundary exported, and new RCR parameters were calculated.

For the present work, no further reduction of M was achieved on the 5th iteration, so the RCR parameters from the 4th iteration were defined as the final parameters, and are listed in Table 3.2. The process took approximately 25–30 h of computational time to obtain the final Windkessel parameters.

Table 3.2 Final tuned Windkessel parameters

	BT	LCC	LS	DA
R_1 (mmHg s/ml)	0.100	0.110	0.150	0.120
R_2 (mmHg s/ml)	2.480	14.590	11.410	2.118
C (ml/mmHg)	0.466	0.085	0.110	0.421

3.2 Methodology

The final simulation (after tuning the parameters) ran for three cycles and took less than eight hours on a standard desktop computer (Quad-core Intel i7 processor, 4 GB RAM). The final cycle was extracted for analysis. The time-step was 5 ms.

Table 3.1 shows that the individual values of the minimum and maximum pressure at each of the outlets are within 5% of the pulse pressure, with the exception of the AA. In the AA, the maximum pressure is very close to the invasively measured value and the mismatch for minimum pressure is within 7% of the pulse pressure. Furthermore, the pulse pressures at each of the outlets are within 2.5%.

Two additional invasive measurements in the distal arch (DAA) and mid-abdominal aorta (MA) in the TL, were compared with the CFD values for further validation. These were found to be in close agreement as shown in Table 3.1.

3.3 Results

3.3.1 Flow Characteristics

Figure 3.2a shows the calculated flow rate through each boundary as a function of time for one cardiac cycle. The peak systolic flow phase ($t = 0.12$ s) is indicated with a vertical dashed line. The LCC and LS received 6.8 and 8.7% of the total inlet flow over the cardiac cycle, respectively, which corresponds to commonly reported values (Tse et al. 2011; Levick 2009). However, 39.4% of the inlet flow went through the BT, leaving only 45.0% of the inlet flow for the DA (both TL and FL). The predicted flow passing through the BT branch shows almost the same wave characteristics as the inlet flow (AA) with a different magnitude. During the negative (diastolic) flow phase, the BT flow rate almost follows that in the AA. All of the aortic branches (BT, LCC and LS) have an early peak phase (phase lead) relative to the inlet flow peak. The peak flow in the DA lags the AA peak and a smaller negative flow passes through

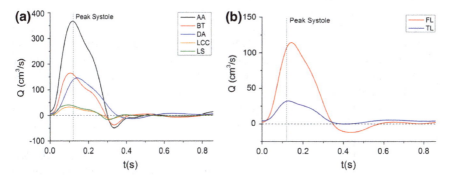

Fig. 3.2 **a** Flow rate at each of the domain boundaries over a single cardiac cycle. **b** Proportion of the flow entering the TL and FL. Vertical *dashed lines* indicate peak systolic flow phase

this outlet compared to the rest of the exit branches. Similar flow wave characteristics can be observed for both LCC and LS branches with a marginally higher magnitude for the LS branch.

Figure 3.2b compares the proportion of the flow through the TL and FL, calculated from the volumetric flow rate through the plane between the two tears indicated in Fig. 2.7 (central horizontal plane). Over the cardiac cycle, around 30% of the flow (which eventually exits via the DA) passed through the TL as opposed to 70% through the FL. Notably, despite the existence of back flow in the DA, there was a negligible amount of reverse flow in the TL, i.e. all of the reverse flow was distributed along the FL. Moreover, the flow in the FL was marginally delayed compared to the TL, as has been observed in clinical studies of AD (Mohr-Kahaly et al. 1989).

Figure 3.3 shows velocity contours along the dissected aorta at different points of the cardiac cycle. Note that the colour bars (contour levels) vary for each sub-figure. Each sub-figure also has an inset with the same contour levels as peak systole (0−2 m/s), in order to aid comparison between time points. At mid-systole (Fig. 3.3a), the velocity varies mainly between 0 and 0.5 m/s along the domain. The velocity in the supraaortic branches is significantly increased due to the decreased cross-sectional area (narrowing) with the highest velocity observed at the LS and LCC branches. At peak-systole (Fig. 3.3b), the velocity increased along the dissected aorta, especially along the coarctation and distal TL downstream of the second tear (re-entry tear). The highest velocity magnitude can be seen in the supraaortic branches. The velocity in the LS branch is significantly skewed towards the wall as a result of the flow in the aneurysm. At the dicrotic notch (Fig. 3.3c), the velocity is still elevated in the branches, with the highest velocity in the LCC. The velocity is slightly increased along the coarctation region at this point. In mid-diastole (Fig. 3.3d), the velocity is relatively uniform throughout the domain ($\approx$0.1 m/s).

The scaled velocity contours in the insets show a velocity increase through the aortic arch branches, distal TL and the coarctation region with the exception of the mid-diastolic phase (Fig. 3.3d). The velocity in the aneurysm and distal and proximal false lumen are very low throughout the cardiac cycle.

Figure 3.4 shows the streamlines obtained at various phases of the cardiac cycle. Relatively uniform streamlines can be observed along the aortic arch and the three branches in mid and peak systolic phases (Fig. 3.4a, b and d). The streamlines in the FL between the two tears are also fairly uniform at peak systole. During the diastolic phase of the cycle (Fig. 3.4c), helical and vortical flow patterns appear, particularly in the FL, and the streamlines become more tortuous throughout the domain. Flow enters the FL via the entry tear and impinges on the wall of the FL (opposite the entry tear), creating chaotic flow patterns. The scaled figures show the streamlines for each phase, with the same contour levels as used for peak systole (Fig. 3.4b) to enable direct comparison.

Figure 3.5 shows streamlines at peak systolic flow for the right anterior and left posterior views of the aneurysm located by the base of the LS. The majority of the flow that enters the aneurysm region exits via the LS branch and the rest recirculates within the aneurysm. The streamlines closest to the posterior wall are reflected and

3.3 Results

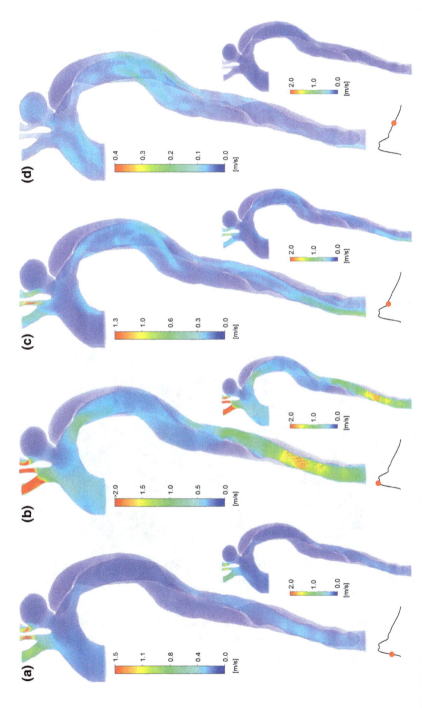

Fig. 3.3 3D velocity contours at **a** mid-systolic pressure phase, **b** peak-systolic pressure phase, **c** dicrotic notch and **d** mid-diastolic pressure phase

78   3 Haemodynamics of a Dissected Aorta

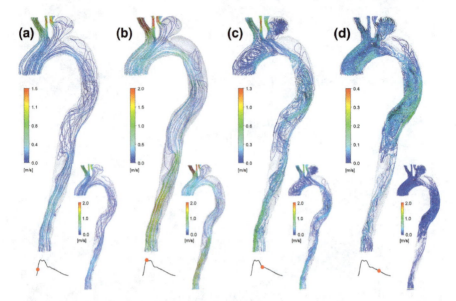

Fig. 3.4 Streamlines at **a** mid-systolic pressure phase, **b** peak-systolic pressure phase, **c** dicrotic notch and **d** mid-diastolic pressure phase

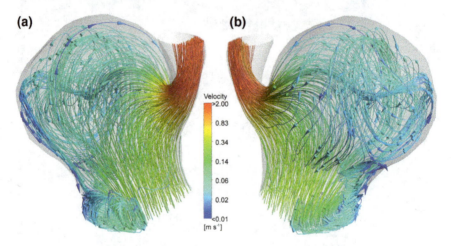

Fig. 3.5 Streamlines in the aneurysm at peak systolic flow (0.12 s) **a** *left* posterior view and **b** *right* anterior view

form a counter-clockwise vortex. In the left posterior view, a well formed vortex in the flow separation zone (at the base of the LS) can also be seen. It is worth noting that the two vortices are counter-rotating.

3.3 Results

Fig. 3.6 Pressure waves at the domain boundaries

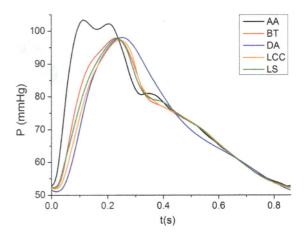

3.3.2 Pressure Distribution

The estimated pressure at the boundaries is shown in Fig. 3.6. It can be seen that the pressure at the outlets lags behind the inlet pressure. The inlet pressure exhibits an uncommon double peak, presumably as a result of the dissection altering the pressure characteristics. This has the appearance of a bisferiens pulse, commonly associated with aortic regurgitation and in some cases aortic stenosis (Ranganathan et al. 2007). It has been reported that the bisferiens pulse occurs as as a result of the 'Venturi effect' (a reduction in pressure due to narrowing of diameter at the stenosed aortic valve (Fleming 1957)). However, as the valve is not included in this model, this cannot explain the observed waveform. Instead, it seems likely that the secondary peak is a result of the reflected wave (Bramwell 1937), occurring due to the restriction caused by the reduced luminal area in the descending aorta. The dicrotic notch is visible in the inlet pressure wave but is entirely absent at the distal abdominal.

Figure 3.7 shows the pressure distribution at four points along the cardiac cycle. As with Fig. 3.3, the contour levels differ for each phase for clarity, and the insets at each phase show the same contours with the colour levels used for peak systole (Fig. 3.7b), for additional comparison.

During the mid-systolic pressure phase (Fig. 3.7a), the pressure drop along the domain is 22 mmHg, with the highest pressure observed along the ascending part of the arch. The pressure decreases towards the outlets. Similarly, in the systolic phase (Fig. 3.7b), the pressure is higher in the TL compared to the same axial location in FL. The opposite relationship is observed for the distal abdominal aorta. The pressure in the aneurysm is high, but is lower than the aortic arch pressure for the same time-instance. Figure 3.7c shows the pressure distribution at the dicrotic notch (during diastole). At this time-point, the pressure distribution is briefly opposite to that observed in the other time-points (Fig. 3.7a, b); the pressure along the aortic arch is lower than in the DA and the pressure at the branch boundaries. This generates the negative flow visible in Fig. 3.2a. The pressure drop across the aorta is about 10 mmHg

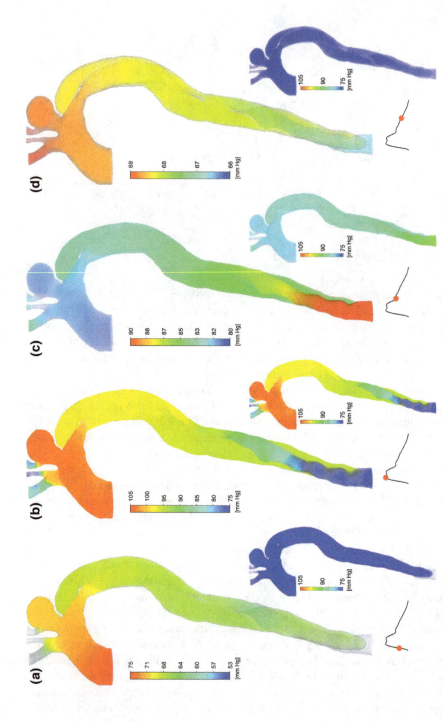

Fig. 3.7 Pressure contours at **a** mid-systolic pressure phase, **b** peak-systolic pressure phase, **c** dicrotic notch and **d** mid-diastolic pressure phase

at this point, with the highest pressure observed in the distal TL (90 mmHg). The pressure along the proximal FL is higher than the same axial location in TL. During the mid-diastolic phase (Fig. 3.7d), the overall pressure distribution is similar to the mid and peak systolic phases (higher pressure along the ascending aorta), the only difference being that the highest pressure is observed at the aortic branches. Additionally, the pressure drop across the domain (AA to DA) is only 3 mmHg; this is the lowest pressure gradient observed between these points, as opposed to the peak systolic phase with the highest pressure drop of 30 mmHg.

3.3.3 Wall Shear Stress

The wall shear stress (WSS) is particularly important in the study of AD, as it affects the structure of the inner wall of the vessel (Gerdes et al. 2000). It is implicated in the initiation stage of AD, as elevated WSS levels can result in cleavage formation (Thubrikar et al. 1999), and subsequently contributes to the growth and enlargement of the FL (Rajagopal et al. 2007). A number of wall shear stress indices are commonly used in the analysis of physiological transient flow simulations in order to describe the WSS characteristics as a single spatial distribution. The time-averaged wall shear stress (TAWSS) is defined by the following expression (Taylor et al. 1999):

$$TAWSS = \frac{1}{T} \int |\tau(t)| \, dt \qquad (3.4)$$

where $|\tau(t)|$ is the magnitude of WSS vector at time t. This equation is applied at each location on the vessel wall to give the TAWSS distribution. An alternative index, intended to provide insight into the nature of the oscillatory forces acting on the endothelium, is the oscillatory shear index (OSI) (Ku et al. 1985; Taylor et al. 1998) described by:

$$OSI = 0.5 \left(1 - \frac{\left| \frac{1}{T} \int_0^T \tau(t) dt \right|}{TAWSS} \right) \qquad (3.5)$$

Figure 3.8a shows the distribution of TAWSS (Eq. 3.4). The TAWSS is at a maximum in the LCC and LS branches, with values of approximately 12 and 8 Pa respectively. This is due to the fact that, despite the relatively low flow rate through the branches, the velocity magnitude is high. In the BT, despite the disproportionately high flow received, the larger diameter reduces the TAWSS compared to the LCC and LS. The TAWSS is also elevated between the coarctation and the entry tear in the TL, and to a lesser extent in both the TL and FL between the tears. Downstream of the re-entry tear, elevated TAWSS values are observed as a result of the significant constriction of the TL.

The magnitude of the TAWSS is very low compared to the pressure (normal stress) shown in Fig. 3.7, (5 Pa is approximately 0.037 mmHg). However, the physiological

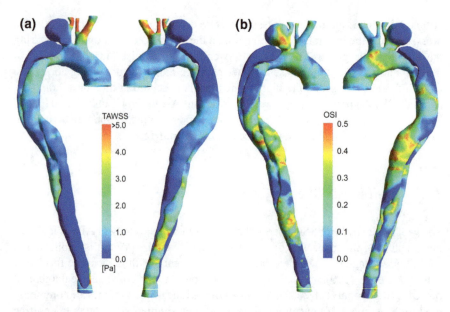

Fig. 3.8 Wall shear stress distributions in the dissected aorta, *left* posterior and *right* anterior views. **a** TAWSS, **b** OSI

relevance of these orthogonal stresses is rather different. Despite the considerably higher normal stresses, the vessel wall structure bears the majority of the stress. However, the endothelial cells lining the vessel wall are highly sensitive to WSS (Van der Heiden et al. 2013), and therefore the role of the shear stress is very important, despite the relatively lower magnitude. In addition, the pressure does also cause expansion of the vessel wall, leading to stretching of the EC, which also induces biological responses changes (Van der Heiden et al. 2013).

The distribution of OSI (Eq. 3.5) is shown in Fig. 3.8b. A region of high OSI can be observed at the aortic arch and only on the posterior side of the aneurysm, in the region where the small strong vortex meets the larger vortex at the wall, as shown in Fig. 3.5a. The OSI was also high in the FL, due to the complex vortical structures present therein, particularly around the tear regions where the interaction with the flow in the TL results in greater oscillations of the flow direction throughout the cardiac cycle. Although the flow between the tears is uniform in the systolic flow phase, it is highly disordered in the diastolic flow phase, as shown in Fig. 3.4b, d.

The regions around the tears are of particular importance, as further tear enlargement could alter the disease state; thus, these regions are considered in more detail in Fig. 3.9, which shows the OSI and streamlines at two points in the cardiac cycle. The time-averaged WSS vectors are also shown in order to help relate the OSI to the preferential flow direction. Considering the entry tear, at peak systole (Fig. 3.9a), the flow upstream of the tear is fairly uniform in the TL, whereas in the FL, vortical flow can be observed, as has been recorded in in vivo studies (Francois et al. 2013).

3.3 Results 83

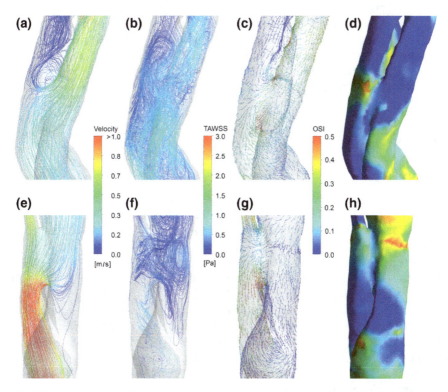

Fig. 3.9 Flow characteristics around the tears at two points in the cardiac cycle. Streamlines at (**a**, **e**) peak systole, (**b**, **f**) mid diastole. **c**, **g** Time-averaged WSS vectors to help relate the OSI to the preferential flow direction (**d**, **h**) OSI. *Top* row shows the entry tear, *bottom* row shows the re-entry tear

Downstream of the tear, the flow is fairly uniform in both lumina (Fig. 3.9a). In mid diastole (Fig. 3.9b), although the mean flow is still visible, the streamlines are highly irregular. The time-averaged WSS vectors appear to be aligned with the flow in the TL upstream of the tear. Correspondingly, the OSI is almost zero in this location, indicating a stable net flow. In the FL, and opposite the tear, a region of very high OSI is visible, which occurs due to separation of the flow as it enters the false lumen. Elevated OSI is observed downstream of the entry tear in both lumina (Fig. 3.9d). The time-averaged WSS vectors also appear highly irregular therein (Fig. 3.9c). Finally, it is worth noting the high magnitude time-averaged WSS vectors at the distal side of the tear. Around the re-entry tear at peak systole (Fig. 3.9e), the streamlines show lower velocity magnitude in the FL than in the TL. Further downstream, the flow velocity in the TL increases substantially due to the merging of the two flow streams from the TL and FL. The FL exhibits negligible flow distal to the re-entry tear. Similarly to the entry tear, streamlines become increasingly disordered as the net flow rate decreases (Fig. 3.9f). In general, the time-averaged WSS vectors are aligned with the flow direction in the TL, except around the re-entry point; however, their directionality varies in the FL, due to the recirculation of the flow before re-entering the TL. The region of highest OSI can thus be observed in the FL, where the main and recirculating flows meet, causing a highly oscillatory force on the wall. Finally, it should be noted that the magnitude of time-averaged WSS vectors is elevated around the distal edge of the tear.

3.4 Discussion

In the present study, 3D simulations of blood flow in a patient-specific AD model have been coupled with Windkessel models at the boundaries. By defining the Windkessel parameters based on patient-specific clinical data, the simulation results take into account the essential characteristics of the rest of the vasculature, which is expected to improve the correlation between the simulation results and the true *in vivo* characteristics. Although newly developed imaging methods, such as 4D MRI, have proved to be successful in measuring the flow (Clough et al. 2012), they still have limited spatial resolution (of the order of 2 mm), which restricts their applicability in calculating WSS parameters, and do not provide pressure information. This is of paramount importance in AD, as it is a pressure related condition. Furthermore, Francois et al. (2013) reported that acquisition of 4D MRI velocity data was not discernible from background noise in the FL for a large proportion of the cardiac cycle.

The present coupling method allows the estimation of the proportion of flow entering each of the branches, the descending aorta, and each of the lumina in the dissected region. Commonly, the proportion of flow entering each of the branches is assumed to be around 5–7.5% (Shahcheraghi et al. 2002; Chen et al. 2013b; Gerdes et al. 2000), on the basis of which more than 75% of the flow would be expected to pass through the descending aorta. However, in type-B ADs, the vascular resistance after the aortic arch increases due to the reduction in the cross sectional area of the TL. As a result, the proportion of flow entering this part of the geometry would be likely to decrease, which might lead to lower limb and/or vital organ (e.g. renal) malperfusion syndrome (Ryan et al. 2013). The simulation results showed a significant reduction in blood supply to the descending aorta, which received only 45% of the total flow. This indicates a risk of lower limb malperfusion which would imply that surgery is necessary (Svensson et al. 2008). If flow proportions were predefined in the simulation using commonly applied values (Shahcheraghi et al. 2002; Tse et al. 2011), then the flow rate through the descending aorta would be significantly higher than for the present simulation (85%). The present results also suggest that the WSS would be overestimated if flow splits were predefined using common values. Conversely, the drastically increased flow in the BT could be expected to significantly increase WSS levels downstream (outside of the computational domain).

The simulations also revealed a region of very high pressure at the point where the BT and LCC bifurcated. The pressure distribution between the two tears was observed to be approximately the same in the FL and TL, possibly due to the relatively short

3.4 Discussion

distance between the two tears. More importantly, the pressure in the FL inferior to the re-entry tear was considerably higher than in the TL at the same transverse location, which in combination with the observed helical flow could lead to further enlargement of the FL (Clough et al. 2012). Potentially, this could in turn lead to occlusion of major arteries and malperfusion. Furthermore, as the thickness of the outer wall of the FL tends to be approximately a quarter of the original thickness of the aortic wall (Shiran et al. 2014), regions of the FL with elevated pressure could be at risk of rupturing, which is the most common cause of death in AD patients (Thubrikar et al. 1999). In addition to the AD, the patient exhibited a large aneurysm at the base of the LS. The predicted pressure values in the aneurysm were as high as those in the ascending aorta and aortic arch. This elevated pressure could potentially lead to further enlargement of the aneurysm. As the aneurysm is very close to the superior part of the FL, and at a high-risk location (Erbel et al. 2001), there is a danger of either aneurysm rupture, or the FL and aneurysm coalescing, with an even higher risk of fatal blowout.

While high WSS is associated with increased risk of tear or aneurysm formation, low WSS values have been observed to correlate with increased risk of expansion within an aneurysm by affecting endothelial functions (Shojima 2004). Studies have shown that, in aneurysms, locations with high values of OSI and low values of TAWSS are more prone to rupture (Xiang et al. 2010). Such a region has been observed in the current results in the LS aneurysm (left-posterior side), due to the interaction of two counter-rotating vortices therein, indicating that this part of the aneurysm is at significant risk of rupture. Additional regions of high TAWSS and low OSI can be observed in the aortic arch and anterior FL adjacent to the entry tear. The former may be a result of the presence of the 'bovine' aortic arch, as patients with this characteristic have an increased risk of thoracic aorta dilation and ascending aorta aneurysm formation, which require surgical intervention (Malone et al. 2012). However, it is not clear whether the correlation of high TAWSS and low OSI also represents a risk within the FL, and this requires further clinical investigation.

As mentioned in the introduction, there is no agreed upon methodology for defining the Windkesssel parameters for the boundary conditions. In the present study, an iterative approach using a simple minimisation algorithm was used. This ensured that appropriate pressure values were predicted at each of the domain boundaries. Although invasive pressure measurements were used for the tuning procedure in this work, by rearranging Eq. 2.15 it would be equally feasible to use, for example, non-invasive ultrasound flow measurements.

3.4.1 Limitations

The input flow wave was not available for the present patient, and hence data from another study was utilised, similarly to the approach employed by Cheng et al. (2010) and Tse et al. (2011), which was selected from a patient with a similar type-B AD. Whilst changes in the characteristics of the flow profile would influence the specific shape of the flow and pressure waves, the tuning procedure computes the pressure waves to emulate the invasive data. However, use of a non patient-specific flow wave may account for some of the mismatch observed between the simulated and measured pressures. This is a limitation of the present study from a patient-specific perspective.

The absence of the visceral arteries in the geometry should also be noted. For the CT scans of the patient provided for this study, the quality and resolution were such that it was not possible to clearly identify these vessels. If they had been included in the model, it would have had a small impact on the quantitative results, but would not significantly effect the overall conclusions, as reported in a previous study (Poelma et al. 2015), who omitted the branches for the sake of efficiency. The flow through the vessels would decrease the amount of flow exiting through the DA, and thus have a small effect on the RCR parameters therein. Furthermore, small regions of modified WSS around the branches would be observed. However, were inclusion of these arteries possible, they would also require appropriate BCs. As no invasive data were available in these branches, the BCs would have to be estimated. The main advantage of being able to include the visceral arteries in the present simulation would have been the additional information that might have been obtained on the possible risk to blockage of these arteries, and thus end-organ malperfusion.

The simulation is based on a rigid wall approximation. Use of the rigid wall approximation has been found to over-predict WSS by 29% compared to fluid-structure interaction (FSI) models, however it does not appear to significantly affect the WSS distribution (Brown et al. 2012; Reymond et al. 2013). It has been reported that patients with AD often have increased aortic stiffness and decreased distensibility (Pyeritz 2000) and that in patients with long term AD, as for the patient in the present study, mobility and elasticity decrease with time as a result of fibrosis (Criado 2011). However, even though the AD walls are reported to be less distensible, the rigid wall assumption fails to capture the 'flutter' motion of the intimal flap between the TL and FL (Ganten et al. 2009). FSI can undoubtedly increase the accuracy of the simulations in theory, however defining the spatially varying Young's modulus and Poisson's ratio is non-trivial (Moireau et al. 2011), particularly for the IF. Given this limitation, results predicted by FSI models are also subject to a degree of uncertainty, and hence there are still improvements to be made in this area in order to develop diagnostic tools based on FSI models. This could potentially be addressed by directly comparing 4D MRI data to computational simulations; however the large cost of MRI scans (Beynon et al. 2012) and limited healthcare resources (Taylor et al. 2005) mean that alternative methods may be necessary. An analysis of the influence of wall motion on AD simulations is carried out in Chap. 5.

Furthermore, FSI simulations are computationally expensive (Brown et al. 2012), making it challenging to embed them in clinical decision support systems, even with the current computational power available. Given the additional computational expense of FSI simulations, the tuning methodology utilised herein should be carried out with rigid wall models. The FSI model could potentially be coupled at the final stage for further tuning if necessary, although this would be a time consuming process.

A possible additional validation for the RCR tuning method would be to build a silicon phantom of the system and couple it with physical Windkessel units (Kung

3.4 Discussion

and Taylor 2010; Kung et al. 2011b). Pressure measurements in the phantom using a catheter could then be used in the tuning methodology, and the outcomes could be compared to the known RCR parameters. However, as the RCR circuit is only an analogue of the downstream vessels, this approach may not actually improve understanding of the model with regards to real clinical data. Acquiring complete invasive pressure measurements (full waves over multiple cardiac cycles) combined with a measurement of the flow or velocity, such as with ultrasound, 2D pcMRI or 4D-MRI, would allow for full validation of the predicted velocity fields and pressures at the domain boundaries.

The present work aimed to complement previous simulation studies on AD in which the BCs (flow or pressure) are prescribed a priori. The use of invasive measurements to help tune RCR parameters with a simple iterative approach provided a methodology that shows promise for future use in the management of AD.

3.5 Sensitivity of the Windkessel Parameters

As the RCR parameter calculation is based on invasive measurements, the sensitivity of the parameters to the measurement resolution of the invasive pressure readings is important. In order to evaluate this, a sensitivity analysis was carried out. The exact location of the invasive pressure measurements cannot be easily discerned, but it is reasonable to assume that the locations reported by the clinician were axially accurate within approximately 1 cm. The pressure drop across such a short distance will be very small and likely within the variability of the pressure measurements. A resolution of ± 1 mmHg is thus a reasonable assumption for the catheter pressure measurements. Hence, the pressure maxima and minima were varied by ± 1 mmHg and corresponding RCR values were calculated.

Figure 3.10 shows the flow waveforms calculated for the plus and minus 1 mmHg cases compared to the original data. In Fig. 3.10a, b, the crosses show the values calculated with the new RCR parameters and the lines show the results analysed in this chapter. Only every fifth point is shown for clarity. Figure 3.10c, d, show the differences between the original and modified data sets. It can be seen that the differences in the flow are negligible for both an increase and a decrease of the pressure by 1 mmHg.

Table 3.3 summarises the calculated Windkessel parameters from the sensitivity analysis as well as the normalised root mean square (RMS) of the difference between the calculated flow rates, given by

$$Q_{RMS}^{*} = \frac{\sqrt{\frac{1}{N}\sum_{n=1}^{N}\left(Q_{s,n} - Q_{o,n}\right)^{2}}}{\frac{1}{N}\sum_{n=1}^{N}Q_{o,n}} \times 100\% \quad (3.6)$$

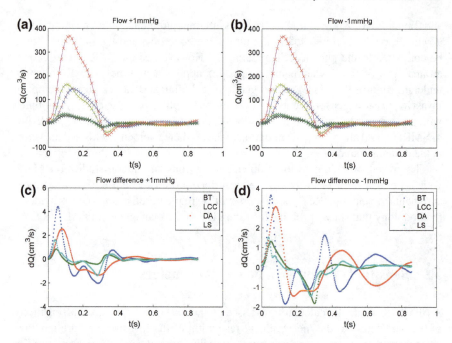

Fig. 3.10 Flow waveforms for **a** $+1$ mmHg, **b** -1 mmHg. Difference in flow rate between the final RCR parameters and **c** $+1$ mmHg, **d** -1 mmHg

Table 3.3 Summary of the Windkessel parameters calculated in the sensitivity analysis. See Eq. 3.6 for definition of Q^*_{RMS}

Windkessel parameters	BT	LCC	LS	DA
R_1 ($+1$ mmHg)	0.100	0.100	0.120	0.160
R_2 ($+1$ mmHg)	2.519	14.800	2.150	11.560
C ($+1$ mmHg)	0.465	0.085	0.420	0.110
Q^*_{RMS} (%) ($+1$ mmHg)	0.044	0.131	0.002	0.126
R_1 (-1 mmHg)	0.090	0.120	0.110	0.140
R_2 (-1 mmHg)	2.457	14.380	2.100	11.250
C (-1 mmHg)	0.460	0.085	0.412	0.110
Q^*_{RMS} (%) (-1 mmHg)	0.081	0.014	0.060	0.086

where n is the index of each timestep and N is the total number of timesteps in one cardiac cycle, Q_o is the original flow rate, and Q_s is the flow rate from the sensitivity analysis.

The small differences show that the selected RCR parameters are relatively insensitive to small changes in the pressure measurements. This provides confidence in the chosen methodology for tuning the RCR parameters.

3.6 Mesh Sensitivity

To assess the mesh sensitivity of this analysis, two additional meshes (a coarse mesh with ≈80,000 and a fine mesh with ≈600,000 elements) were created with the same meshing configuration that was described in Chap. 2 (tetrahedral elements with prismatic wall layers).

There are no strict guidelines for the number of mesh elements required to achieve a sufficiently accurate solution. Indeed, the number is highly dependent on the geometry and the type of solver used. In order to establish whether a chosen grid is sufficiently fine, it is therefore common practice to evaluate the differences in key variables if relatively coarse or fine meshes were used, and to compare relative gains in accuracy, to increases in computational expense. In the context of simulations blood flow in aortae, a wide range of mesh cardinalities have been reported in idealised aortic geometries; 70,000 (Shahcheraghi et al. 2002), 80,000 (Wen et al. 2009), 130,000 (Lagana et al. 2005; Migliavacca et al. 2006) 350,000 (Benim et al. 2011) and in healthy aortic geometries from patients; 110,000 (Moireau et al. 2013), 380,000 (Reymond et al. 2013), 500,000 (Brown et al. 2012). In AD simulations, Karmonik et al. (2012a) used 190,000 and 250,000 for initial and follow-up geometries, while Tse et al. (2011) used 115,000 and 84,000 for pre- and post-aneurysm geometries. However, other groups have used cell counts from 1 to 3 million for AD simulations (Chen et al. 2013b, a; Cheng et al. 2014, 2010). In the present study, as clinical translation is one of the stated long-term aims, computational efficiency is paramount, provided that the solution is relatively insensitive to further refinements in the mesh. If significant changes between medium and fine meshes were observed, then further mesh refinements would be required. In the following section, the key results (flows and pressures at the boundaries, wall shear indices, and pressure and velocity distributions) are compared in detail between the three meshes.

Sample images of the three meshes can be found in Appendix A. The same BCs were applied at the boundaries, creating the same environment (RCR parameters and input flow) as the one employed for the results described in this chapter (medium mesh).

Figure 3.11a shows that the volumetric flow rate is the same in the AA for all three meshes, as it is prescribed. However, the inlet pressure (3.11b) for the coarse mesh differs significantly from that obtained with the fine and medium meshes, resulting in a value that is significantly different to the invasive pressure measurements.

Figure 3.12a compares the volumetric flow rate between the three mesh cases in the BT. The coarse mesh shows a slightly higher value of the volumetric flow rate which also results in a slight reduction in the pressure (Fig. 3.12b). Overall, all three cases are similar for both flow and pressure values.

The volumetric flow rate and pressure for the LCC branch are shown in Fig. 3.13a, b. The volumetric flow rates for the medium and fine meshes are very similar. However, differences can be seen for the coarse mesh results. Similarly, the maximum pressure (Fig. 3.13b) for the coarse mesh is marginally lower than for the medium and fine meshes.

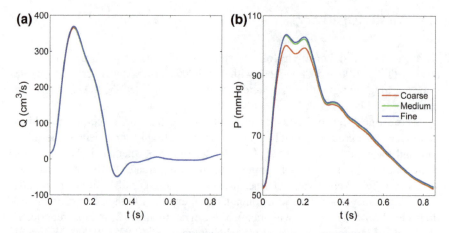

Fig. 3.11 a Flow at the AA for all three meshes. b Pressure at the AA for all three meshes

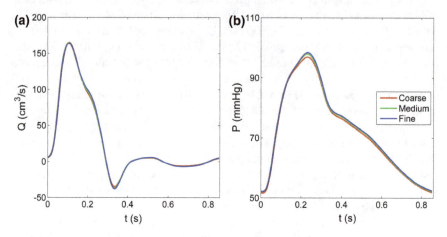

Fig. 3.12 a Flow at the BT for all three meshes. b Pressure at the BT for all three meshes

For the LS branch (Fig. 3.14), the volumetric flow rate (Fig. 3.14a) for the coarse mesh has a smaller value at peak systole and less negative flow at $t = 0.3$ s. The pressure waveform for the coarse mesh deviates from the medium and fine meshes, particularly at peak systole. The pressure and flow for the medium and fine meshes are very similar.

Figure 3.15a shows the volumetric flow rate in all cases for the DA. The medium and fine meshes have similar characteristics. The volumetric flow rate wave for the coarse mesh at the DA differs only slightly from the other two models. There is no discernible difference between the meshes for the pressure at the DA Fig. 3.15b.

It should be noted that the minimum pressures at the boundaries for all three mesh cases in all branches are the same, and the waveforms vary only significantly around systole.

3.6 Mesh Sensitivity

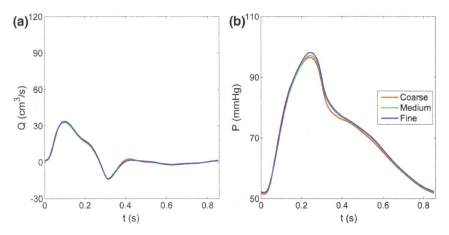

Fig. 3.13 a Flow at the LCC for all three meshes. **b** Pressure at the LCC for all three meshes

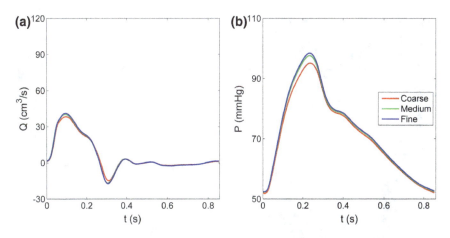

Fig. 3.14 a Flow at the LS for all three meshes. **b** Pressure at the LS for all three meshes

Figures 3.16a, 3.17a and 3.18a show the TAWSS in the left posterior and right anterior views for the three mesh cases, respectively. In all cases, the maximum TAWSS can be seen through the aortic arch branches (BT, LCC and LS), coarctation and the distal TL; however, the magnitudes are different. In the coarse mesh, the highest value is ≈8 Pa, compared to ≈12 Pa in both medium and fine mesh cases (although this cannot be seen in the figure, as the colour scales show only up to 5 Pa for clarity). The TAWSS in the medium and fine mesh are similar in magnitude and distribution. The only differences between these two models are the slight increase in TAWSS values in the aortic arch and distal TL for the fine mesh compared to the medium mesh.

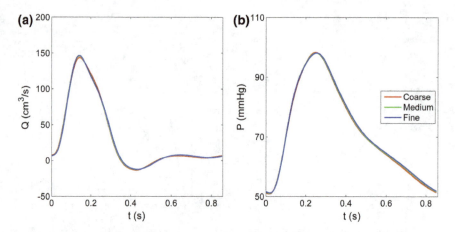

Fig. 3.15 **a** Flow at the DA for all three meshes. **b** Pressure at the DA for all three meshes

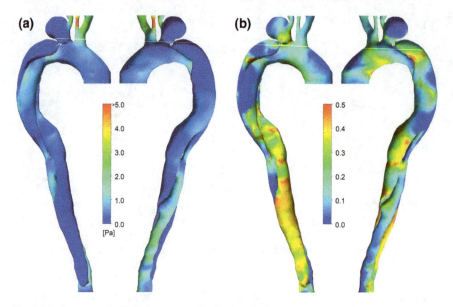

Fig. 3.16 Coarse mesh: Wall shear stress distributions in the aorta showing *left* posterior and *right* anterior views. **a** TAWSS, **b** OSI

Figures 3.16b, 3.17b and 3.18b show the distribution of OSI for all three mesh cases. The OSI for the coarse mesh differs considerably from the other two mesh cases (medium and fine) in both magnitude and distribution.

On the contrary, the OSI distribution is very similar for the medium and fine meshes. The OSI in the left posterior view of the aneurysm (by the LS branch) was decreased for the fine mesh compared to the same location in the medium mesh and was slightly increased (by 0.1) in the right anterior view of the aneurysm. The OSI

3.6 Mesh Sensitivity

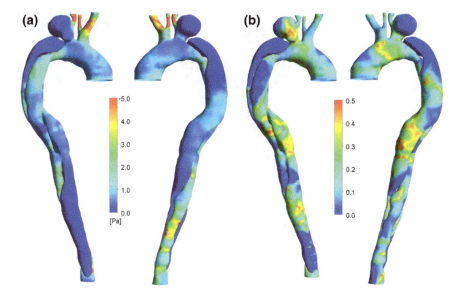

Fig. 3.17 Medium mesh: Wall shear stress distributions in the aorta showing *left* posterior and *right* anterior views. **a** TAWSS, **b** OSI

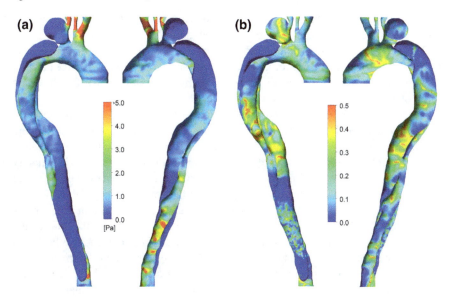

Fig. 3.18 Fine mesh: Wall shear stress distributions in the aorta showing *left* posterior and *right* anterior views. **a** TAWSS, **b** OSI

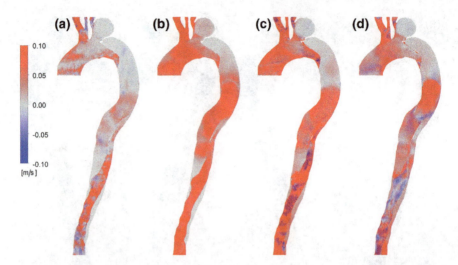

Fig. 3.19 Velocity differences between the medium and fine meshes at **a** peak systole and **b** dicrotic notch. Velocity differences between the coarse and medium meshes at **c** peak systole and **d** dicrotic notch

was also increased in the proximal FL and decreased in the distal FL compared to the medium mesh in the same locations, although the differences are small.

The peak systolic and the dicrotic notch time points were selected to compare the difference between estimated velocities and pressures obtained with the three meshes. The differences were calculated by importing both results files into CFX-Post, which interpolated the data onto the coarser mesh for each case.

The difference in the predicted velocities between the medium and fine meshes, and coarse and medium meshes are shown in Fig. 3.19. Comparing the medium and fine meshes at peak systole (3.19a), the largest velocity difference is ≈0.1 m/s along the aortic arch, BT and in between the two tears in the FL and distal TL. The velocity differences are larger at the dicrotic notch (3.19b) throughout the domain, with the exception of the proximal and distal FL, wherein the velocity is negligible.

At peak systole (Fig. 3.19c), there is a considerable difference in the local velocities obtained with the medium and coarse meshes, particularly at the wall of the arch. The relatively larger prism layer size in the coarse mesh results in a failure to adequately resolve the flow, and so the velocity is significantly underestimated. This occurs throughout the domain, resulting in the decreased values of TAWSS shown in Fig. 3.16b. At the dicrotic notch (Fig. 3.19d) the differences are similar.

Figure 3.20 shows the pressure differences between the medium and fine meshes (Fig. 3.20a, b), and coarse and medium meshes (Fig. 3.20c, d) at two critical time points in the cardiac cycle. The largest pressure difference (−4 mmHg) can be observed at peak systole (Fig. 3.20c) between the coarse and medium meshes throughout the aorta, except the distal TL and the LCC branch. At the dicrotic notch (Fig. 3.20d), the differences between the medium and coarse meshes are negligible

3.6 Mesh Sensitivity

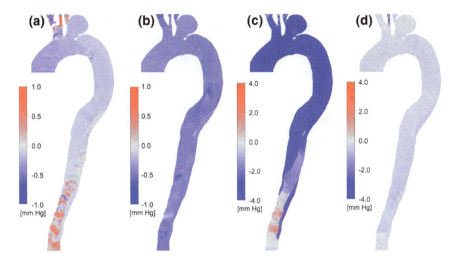

Fig. 3.20 Pressure differences between the medium and fine meshes at **a** peak systole and **b** dicrotic notch. Pressure differences between the coarse and medium meshes at **c** peak systole and **d** dicrotic notch

throughout the domain. The difference in pressure between the medium and fine meshes is small; ±1 mmHg for both time points shown (Fig. 3.20a, b).

These comparisons were performed to select the optimum mesh in terms of both computational time (simulation time until it reaches periodicity) and performance (accuracy of the results). Based on this analysis, the coarse mesh could not be used due to insufficient accuracy (mesh size sensitivity) of the results, despite being considerably faster than the medium and fine meshes. The fine mesh produced results that were comparable to the medium mesh; however, the computational time was increased by a factor of 3. Thus, the medium mesh was deemed sufficient to be used for this patient-specific study, as the flow and pressure waveforms at the BCs and the haemodynamic parameters along the dissected region did not vary significantly when compared to the fine case.

3.7 Effect of Turbulence Modelling

Finally, in this section the possible impact of the choice of a laminar flow model is investigated. In order to examine the influence of turbulence on the present simulations, the shear stress transport (SST) model, a version of Menter's hybrid $k-\varepsilon/k-\omega$ model, was implemented in CFX, following the same methodology as Chen et al. (2013a). The same boundary conditions were applied as for the laminar model, with a 1% turbulence intensity (Chen et al. 2013a; Tan et al. 2009) applied at the boundaries using the medium mesh.

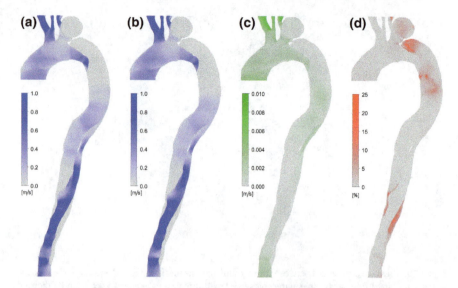

Fig. 3.21 At peak systole: **a** Velocity for the laminar flow model, **b** velocity for the turbulent model, **c** magnitude of the turbulent velocity fluctuations, **d** turbulence intensity

Figure 3.21 shows the velocity, the turbulent velocity fluctuations, and the turbulence intensity at peak systole. Comparison of the velocities obtained with the laminar and turbulent models (Fig. 3.21a, b respectively), shows no significant differences between the two. The magnitude of the velocity fluctuations is shown in Fig. 3.21c. It can be seen that the regions of elevated turbulent velocity fluctuations correspond approximately to those of higher overall velocity, although the magnitude is very small (<0.01 m/s: note the different colour scales compared to the overall velocity). The turbulence intensity distribution (Fig. 3.21d) shows several regions of turbulence up to 25%; however it can be seen that these occur predominantly in the proximal and distal FL. Considering both Fig. 3.21b, c, it is clear that the values of elevated Tu are the result of the low local velocity in these regions, rather than a significant amount of turbulence fluctuation, and thus have very little turbulent energy associated with them. As a result, turbulence in these regions is not expected to influence the results.

Hence, it can be concluded that turbulence did not significantly affect the results of the present AD simulation. In order to justify whether it is appropriate to use a turbulence model in this context, the additional computational time to run such simulations should be considered. For the mesh used, each time step took roughly twice as long for the turbulent model compared to the laminar model. The laminar model appears to be an appropriate simplification, as it halved the computational cost without significant impact on the results. The negligible difference between the turbulent and laminar simulation results can be ascribed to the fact that what little turbulence was observed had very little energy associated with it. Therefore, the velocity predicted by the two models was practically identical. However, in the context of other geometries or in more complex simulations, such as using fluid-structure interaction, the influence of turbulence modelling may need to be re-evaluated.

3.8 Conclusions

The application of patient-specific boundary conditions to a CFD simulation of blood flow in a type-B communicating dissected aorta has been reported in this chapter. A novel tuning methodology to define the parameters of the Windkessel models coupled to the boundaries, based on the maxima and minima of invasive pressure measurements, has been developed and investigated. This approach provided predictions of fluid dynamics in the dissected aorta with high resolution. The predicted pressure values in the false lumen are expected to be more accurate in absolute terms than in simulations with prescribed flow or pressure waves. Furthermore, as organ ischemia is present in many AD patients, correct approximation of blood flow through each region affects the predicted haemodynamic factors and should be considered in AD studies. For the present patient, turbulence was not found to significantly affect the results.

The present chapter represents a step towards the long-term aim of embedding CFD simulations in the clinic as a tool to provide additional information for clinicians (Taylor et al. 1999). Although further development is required, this work has shown that the tuning of RCR parameters based on clinical data is a viable option in order to provide patient-specific boundary conditions in the analysis of AD.

In the next chapter, an application of patient-specific simulations as a predictive tool for intervention and decision making is described. Virtual stenting operations are performed on the same geometry, in order to evaluate the haemodynamic efficacy of different treatment options.

References

Alimohammadi, M., Agu, O., Balabani, S., & Díaz-Zuccarini, V. (2014a). Development of a patient-specific tool to analyse aortic dissections: Assessment of pamixed patient-specific flow and pressure boundary conditions. *Medical Engineering and Physics, 36*, 275–284.

Alimohammadi, M., Pichardo-Almarza, C., Di Tomaso, G., Balabani, S., Agu, O., & Díaz-Zuccarini, V. (2015). Predicting atherosclerotic plaque location in an iliac bifurcation using a hybrid CFD/Biomechanical approach, in IWBBIO 2015. *LNCS, 9044*, 594–606.

Benim, A. C., Nahavandi, A., Assmann, A., Schubert, D., Feindt, P., & Suh, S. H. (2011). Simulation of blood flow in human aorta with emphasis on outlet boundary conditions. *Applied Mathematical Modelling, 35*, 3175–3188.

Bertoglio, C., Moireau, P., & Gerbeau, J.-F. (2011). Sequential parameter estimation for fluid-structure problems: Application to hemodynamics. *International Journal for Numerical Methods in Biomedical Engineering, 28*(4), 434–455.

Beynon, R., Sterne, J. A. C., Wilcock, G., Likeman, M., Harbord, R. M., Astin, M., et al. (2012). Is MRI better than CT for detecting a vascular component to dementia? a systematic review and meta-analysis. *BMC Neurology, 12*(1), 33.

Blanco, P. J., Feijóo, R. A., & Urquiza, S. A. (2007). A unified variational approach for coupling 3D–1D models and its blood flow applications. *Computer Methods in Applied Mechanics and Engineering, 196*(41–44), 4391–4410.

Bramwell, C. (1937). The arterial pulse in health and disease. *The Lancet, 230*, 301–305.

Brown, A. G., Shi, Y., Marzo, A., Staicu, C., Valverde, I., Beerbaum, P., et al. (2012). Accuracy versus computational time translating aortic simulations to the clinic. *Journal of Biomechanics, 45*(3), 516–523.

Chen, D., ller Eschner, M. M., von Tengg-Kobligk, H., Barber, D., Bockler, D., Hose, R., et al. (2013a). A patient-specific study of type-B aortic dissection: Evaluation of true-false lumen blood exchange. *BioMedical Engineering OnLine, 12*, 65.

Cheng, Z., Juli, C., Wood, N. B., Gibbs, R. G. J., & Xu, X. Y. (2014). Predicting flow in aortic dissection: Comparison of computational model with PC-MRI velocity measurements. *Medical Engineering and Physics, 36*(9), 1176–1184.

Cheng, Z., Riga, C., Chan, J., Hamady, M., Wood, N. B., Cheshire, N. J., et al. (2013). Initial findings and potential applicability of computational simulation of the aorta in acute type B dissection. *Journal of Vacscular Surgery, 57*(2), 35S–43S.

Cheng, Z., Tan, F. P. P., Riga, C. V., Bicknell, C. D., Hamady, M. S., Gibbs, R. G. J., et al. (2010). Analysis of flow patterns in a patient-specific aortic dissection model. *Journal of Biomechanical Engineering, 132*(5), 051007.

Chen, D., Müller-Eschner, M., Kotelis, D., Böckler, D., Ventikos, Y., & von Tengg-Kobligk, H. (2013b). A longitudinal study of Type-B aortic dissection and endovascular repair scenarios: Computational analyses. *Medical Engineering and Physics, 35*(9), 1321–1330.

Clough, R. E., Waltham, M., Giese, D., Taylor, P. R., & Schaeffter, T. (2012). A new imaging method for assessment of aortic dissection using four-dimensional phase contrast magnetic resonance imaging. *Journal of Vacscular Surgery, 55*(4), 914–923.

Criado, F. J. (2011). Aortic dissection: A 250-year perspective. *Texas Heart Institute Journal, 38*(6), 694–700.

D'Elia, M., Perego, M., & Veneziani, A. (2011). A variational data assimilation procedure for the incompressible Navier–Stokes Equations in hemodynamics. *Journal of Scientific Computing, 52*(2), 340–359.

Erbel, R., Alfonso, F., Boileau, C., Dirsch, O., Eber, B., Haverich, A., et al. (2001). Diagnosis and management of aortic dissection task force on aortic dissection, European society of cardiology. *European Heart Journal, 22*(18), 1642–1681.

Fattori, R., Cao, P., De Rango, P., Czerny, M., Evangelista, A., Nienaber, C., et al. (2013). Interdisciplinary expert consensus document on management of type B aortic dissection. *Journal of the American College of Cardiology, 61*(16), 1661–1678.

Fleming, P. R. (1957). The mechanism of the pulsus bisferiens. *British heart journal, 19*(4), 519–524.

Formaggia, L., Nobile, F., Quarteroni, A., & Veneziani, A. (1999). Multiscale modelling of the circulatory system: A preliminary analysis. *Computing and Visualization in Science, 2*, 75–83.

Formaggia, L., Quarteroni, A. M., & Veneziani, A. (2009). *Cardiovascular mathematics: Modeling and simulation of the circulatory system*. New York: Springer.

Francois, C. J., Markl, M., Schiebler, M. L., Niespodzany, E., Landgraf, B. R., Schlensak, C., et al. (2013). Four-dimensional, flow-sensitive magnetic resonance imaging of blood flow patterns in thoracic aortic dissections. *The Journal of Thoracic and Cardiovascular Surgery, 145*(5), 1359–1366.

Ganten, M.-K., Weber, T. F., von Tengg-Kobligk, H., Böckler, D., Stiller, W., Geisbüsch, P., et al. (2009). Motion characterization of aortic wall and intimal flap by ECG-gated CT in patients with chronic B-dissection. *European Journal of Radiology, 72*(1), 146–153.

Gerdes, A., Joubert-Hübner, E., Esders, K., & Sievers, H. H. (2000). Hydrodynamics of aortic arch vessels during perfusion through the right subclavian artery. *The Annals of Thoracic Surgery, 69*(5), 1425–1430.

Karmonik, C., Bismuth, J. X., Davies, M. G., & Lumsden, A. B. (2008). Computational hemodynamics in the human aorta: A computational fluid dynamics study of three cases with patient-specific geometries and inflow rates. *Technology and Health Care, 16*(5), 343–354.

References

Karmonik, C., Bismuth, J., Shah, D. J., Davies, M. G., Purdy, D., & Lumsden, A. B. (2011b). Computational study of haemodynamic effects of entry- and exit-tear coverage in a DeBakey type III aortic dissection: Technical report. *European Journal of Vascular and Endovascular Surgery, 42*(2), 172–177.

Karmonik, C., Duran, C., Shah, D. J., Anaya-Ayala, J. E., Davies, M. G., Lumsden, A. B., et al. (2012a). Preliminary findings in quantification of changes in septal motion during follow-up of type B aortic dissections. *Journal of Vacscular Surgery, 55*(5), 1419–1426.e1.

Kim, H. J., Vignon-Clementel, I. E., Figueroa, C. A., LaDisa, J. F., Jansen, K. E., Feinstein, J. A., et al. (2009). On coupling a lumped parameter heart model and a three-dimensional finite element aorta model. *Annals of Biomedical Engineering, 37*(11), 2153–2169.

Ku, D. N., Giddens, D. P., Zarins, C. K., & Glagov, S. (1985). Pulsatile flow and atherosclerosis in the human carotid bifurcation: Positive correlation between plaque location and low oscillating shear stress. *Arteriosclerosis Thrombosis and Vascular Biology, 5*(3), 293–302.

Kung, E. O., Les, A. S., Medina, F., Wicker, R. B., McConnell, M. V., & Taylor, C. A. (2011b). In vitro validation of finite-element model of AAA hemodynamics incorporating realistic outlet boundary conditions. *Journal of Biomechanical Engineering, 133*(4), 041003.

Kung, E. O., & Taylor, C. A. (2010). Development of a physical windkessel module to re-create in vivo vascular flow impedance for in vitro experiments. *Cardiovascular Engineering and Technology, 2*(1), 2–14.

Laganà, K., Balossino, R., Migliavacca, F., Pennati, G., Bove, E. L., de Leval, M. R., et al. (2005). Multiscale modeling of the cardiovascular system: Application to the study of pulmonary and coronary perfusions in the univentricular circulation. *Journal of Biomechanics, 38*, 1129–1141.

Levick, J. (2009). *An introduction to cardiovascular physiology* (5th ed.). London: Hodder Arnold.

Malone, C. D., Urbania, T. H., Crook, S. E. S., & Hope, M. D. (2012). Bovine aortic arch: A novel association with thoracic aortic dilation. *Clinical Radiology, 67*(1), 28–31.

Migliavacca, F., Balossino, R., Pennati, G., Dubini, G., Hsia, T.-Y., de Leval, M. R., et al. (2006). Multiscale modelling in biofluidynamics: Application to reconstructive paediatric cardiac surgery. *Journal of Biomechanics, 39*(6), 1010–1020.

Mohr-Kahaly, S., Erbel, R., Rennollet, H., Wittlich, N., Drexler, M., Oelert, H., et al. (1989). Ambulatory follow-up of aortic dissection by transesophageal two- dimensional and color-coded doppler echocardiography. *Circulation, 80*(1), 24–33.

Moireau, P., Bertoglio, C., Xiao, N., Figueroa, C. A., Taylor, C. A., Chapelle, D., et al. (2013). Sequential identification of boundary support parameters in a fluid-structure vascular model using patient image data. *Biomechanics and Modeling in Mechanobiology, 12*(3), 475–496.

Moireau, P., Xiao, N., Astorino, M., Figueroa, C. A., Chapelle, D., Taylor, C. A., et al. (2011). External tissue support and fluid–structure simulation in blood flows. *Biomechanics and Modeling in Mechanobiology, 11*, 1–18.

Poelma, C., Watton, P. N., & Ventikos, Y. (2015). Transitional flow in aneurysms and the computation of haemodynamic parameters. *Journal of The Royal Society Interface, 12*(105), 20141394.

Pyeritz, R. E. (2000). The marfan syndrome. *Annual Review of Medicine, 51*, 481–510.

Quarteroni, A., Veneziani, A., & Zunino, P. (2002). Mathematical and numerical modeling of solute dynamics in blood flow and arterial walls. *SIAM Journal on Numerical Analysis, 39*(5), 1488–1511.

Rajagopal, K., Bridges, C., & Rajagopal, K. R. (2007). Towards an understanding of the mechanics underlying aortic dissection. *Biomechanics and Modeling in Mechanobiology, 6*(5), 345–359.

Ranganathan, N., Sivaciyan, V., & Saksena, F. B. (2007). *The art and science of cardiac physical examination: With heart sounds and pulse wave forms on CD*. New York: Springer.

Reymond, P., Crosetto, P., Deparis, S., Quarteroni, A., & Stergiopulos, N. (2013). Physiological simulation of blood flow in the aorta: Comparison of hemodynamic indices as predicted by 3-D FSI, 3-D rigid wall and 1-D models. *Medical Engineering and Physics, 35*(6), 784–791.

Ryan, C., Vargas, L., Mastracci, T., Srivastava, S., Eagleton, M., Kelso, R., et al. (2013). Progress in management of malperfusion syndrome from type B dissections. *Journal of Vacscular Surgery, 57*(5), 1283–1290.

Shahcheraghi, N., Dwyer, H. A., Cheer, A. Y., Barakat, A. I., & Rutaganira, T. (2002). Unsteady and three-dimensional simulation of blood flow in the human aortic arch. *Journal of Biomechanical Engineering, 124*(4), 378.

Shi, Y., Lawford, P., & Hose, R. (2011). Review of zero-D and 1-D models of blood flow in the cardiovascular system. *BioMedical Engineering OnLine, 10*(1), 33.

Shiran, H., Odegaard, J., Berry, G., Miller, D. C., Fischbein, M., & Liang, D. (2014). Aortic wall thickness: An independent risk factor for aortic dissection? *The Journal of heart valve disease, 23*(1), 17–24.

Shojima, M. (2004). Magnitude and role of wall shear stress on cerebral aneurysm: Computational fluid dynamic study of 20 middle cerebral artery aneurysms. *Stroke, 35*(11), 2500–2505.

Suh, G.-Y., Les, A. S., Tenforde, A. S., Shadden, S. C., Spilker, R. L., Yeung, J. J., et al. (2010). Quantification of particle residence time in abdominal aortic aneurysms using magnetic resonance imaging and computational fluid dynamics. *Annals of Biomedical Engineering, 39*(2), 864–883.

Svensson, L. G., Kouchoukos, N. T., Miller, D. C., Bavaria, J. E., Coselli, J. S., Curi, M. A., et al. (2008). Expert consensus document on the treatment of descending thoracic aortic disease using endovascular stent-grafts. *The Annals of Thoracic Surgery, 85*(1), S1–S41.

Swee, W., & Dake, M. D. (2008). Endovascular management of thoracic dissections. *Circulation, 117*, 1460–1473.

Tan, F. P. P., Borghi, A., Mohiaddin, R. H., Wood, N. B., Thom, S., & Xu, X. Y. (2009). Analysis of flow patterns in a patient-specific thoracic aortic aneurysm model. *Computers and Structures, 87*(11–12), 680–690.

Taylor, C., Draney, M., Ku, J., Parker, D., Steele, B., Wang, K., et al. (1999). Predictive medicine: Computational techniques in therapeutic decision-making. *Computer Aided Surgery, 4*, 231–247.

Taylor, C. A., Hughes, T. J. R., & Zarins, C. K. (1998). Finite element modeling of three-dimensional pulsatile flow in the abdominal aorta: Relevance to atherosclerosis. *Annals of Biomedical Engineering, 26*(6), 975–987.

Taylor, A., Sheridan, M., McGee, S., & Halligan, S. (2005). Preoperative staging of rectal cancer by MRI; results of a UK survey. *Clinical Radiology, 60*, 579–586.

Thubrikar, M. J., Agali, P., & Robicsek, F. (1999). Wall stress as a possible mechanism for the development of transverse intimal tears in aortic dissections. *Journal of Medical Engineering & Technology, 23*(4), 127–134.

Tse, K. M., Chiu, P., Lee, H. P., & Ho, P. (2011). Investigation of hemodynamics in the development of dissecting aneurysm within patient-specific dissecting aneurismal aortas using computational fluid dynamics (CFD) simulations. *Journal of Biomechanics, 44*(5), 827–836.

Van der Heiden, K., Gijsen, F. J. H., Narracott, A., Hsiao, S., Halliday, I., Gunn, J., et al. (2013). The effects of stenting on shear stress: Relevance to endothelial injury and repair. *Cardiovascular Research, 99*(2), 269–275.

Wen, C.-Y., Yang, A.-S., Tseng, L.-Y., & Chai, J.-W. (2009). Investigation of pulsatile flowfield in healthy thoracic aorta models. *Annals of Biomedical Engineering, 38*(2), 391–402.

Xiang, J., Natarajan, S. K., Tremmel, M., Ma, D., Mocco, J., Hopkins, L. N., et al. (2010). Hemodynamic-morphologic discriminants for intracranial aneurysm rupture. *Stroke, 42*(1), 144–152.

Chapter 4
Effectiveness of Aortic Dissection Treatments via Virtual Stenting

In this chapter, pre-operative and virtual-stenting scenarios are applied to the same patient model that was discussed in the previous chapter. The LS aneurysm is removed for the purpose of simplicity in order to yield more general results. The 3D models are coupled to the same Windkessel models used in the previous chapter and for each case, velocity, pressure, WSS and energy loss are evaluated and compared. The methodology applied in the present study reveals detailed information about two possible surgical interventions and shows promise as a diagnostic and interventional planning tool.

It should be noted that the term 'virtual-stenting' is used in this chapter to mean 'virtually simulating the effect of a stenting operation on the anatomy of a dissected aorta', as opposed to directly simulating the process of deploying the stent within the artery, as has been reported in a number of recent studies (Spranger and Ventikos 2015; Spranger et al. 2015; Chen et al. 2013c).

4.1 Introduction

In the previous chapter, the implementation of dynamic BCs (tuned according to the maxima and minima of invasive pressure measurements) enabled acquisition of high-resolution, interdependent, time-varying flow and pressure distributions for each boundary. Patients with type-B AD often suffer from a range of other vascular-related conditions as a result of the dissection. These complications can be end-organ malperfusion (organ ischaemia), major branch obstruction, hypertension, aneurysm

---

The work presented in this chapter was published in 'Evaluation of the haemodynamic effectiveness of aortic dissection treatments via virtual stenting', *International Journal of Artificial Organs* (Alimohammadi et al. 2014b).

formation, rupture of the vessel wall etc. (Criado 2011; Fattori et al. 2008b; Hagan et al. 2000; Svensson et al. 2008). This makes the condition highly specific to each patient, which emphasises the importance of patient-tailored treatments in the case of AD.

There is no single protocol regarding the treatment of type-B patients, as each patient requires treatment based on their specific conditions (also known as patient-tailored therapy) (Tsai et al. 2008; Umaña et al. 2002). In the presence of complications in cases of type-B AD, open surgery or endovascular treatment may become necessary (Nordon et al. 2011). However, open surgery has significant risks such as early mortality, stroke and spinal cord ischaemia (Fattori et al. 2013).

One of the most common surgical treatments in the management of type-B AD is thoracic endovascular aortic repair (TEVAR). In this procedure, a collapsed stent is inserted into the aorta via a small incision in the lower limb and is expanded *in situ*. This method is minimally invasive and does not require extensive open surgery, however the post-operative complications of the method are not yet clear (Svensson et al. 2008; Grabenwoger et al. 2012). In a complicated AD, TEVAR can be used to minimise further distension of the vessel wall and to reduce the effect of malperfusion (Hagan et al. 2000; Nordon et al. 2011). One of the most significant advantages of using TEVAR, as opposed to open surgery, is the decreased morbidity, mortality and patient recovery length (Gopaldas et al. 2010), in addition to the safety and effectiveness of the method (Nienaber et al. 1999). Graft migration, delayed complications and endoleak have been reported as short term implications, but due to the surgery being relatively new, long term results are not presently available (Gopaldas et al. 2010; McPhee et al. 2007).

Selection of the optimal placement for stent-graft treatment is a matter of importance. The least invasive surgical option is to place the stent-graft so as to occlude the entry tear, restricting the flow of blood in the FL in order to induce vascular remodelling (Nienaber 2011), and hence expansion of the TL and retraction of the FL (van Bogerijen et al. 2014). Sayer et al. (2008) found that in acute type-B dissections, the FL diameter reduced to zero after 12 months and the TL diameter more than doubled at the site of the stent-graft. However, distal to the stent-graft the relative changes were less consistent, suggesting the need for longer stents to cover the entire descending thoracic aorta (Kölbel et al. 2014). However, the need to ensure blood flow through the visceral arteries requires the use of fenestrated grafts, in cases where the tears extend below these arteries. Fenestrated grafts are an established technique in thoracic pathologies involving the visceral arteries (Scali et al. 2013b; Verhoeven et al. 2012). Kitagawa et al. (2013) applied this technique, which uses grafts with fenestrations for the visceral arteries, to type-B AD patients. They reported that extending the graft both distal and proximal to the dissection was successful in obliterating the false lumen. This technique is technically challenging, but shows great promise (Kölbel et al. 2014). Song et al. (2014) found that complete coverage of all re-entry tears was successful in patients with chronic type-B AD, but is limited to tears located above the celiac artery. Sayer et al. (2008) reported that FL thrombosis was not achieved after covering the proximal entry tear, and thus additional procedures were required. They concluded that their previous policy of occluding the entry tear and monitoring for complications needed review, in the context of two

4.1 Introduction

patients who had rupture of the FL. Techniques such as blocking the FL with filters, balloons, coils or glue have been reported (Loubert et al. 2003; Hofferberth et al. 2012). Sayer et al. (2008) propose treating the entire thoracic aorta (including the distal regions) at the same time as the initial TEVAR operation with debranching and endovascular repair.

Due to the complexities of the various treatment options, it is important to be able to provide detailed, patient-specific, predictions of the effects of different surgical procedures, so as to provide the clinician with all the necessary information required in the decision making process. In the case of TEVAR, CFD could provide a detailed prediction of clinically relevant haemodynamic parameters pre- and post-stenting, in order to aid the decision making process regarding whether to surgically intervene (Karmonik et al. 2011b; Chen et al. 2013b).

Karmonik et al. (2011b) carried out CFD simulations in a type-B dissected aorta and analysed the effect of occluding the entry tear, re-entry tear, or both. They reported occluding the entry tear as a prediction of standard TEVAR treatment, whilst occluding both tears in their study was a model for surgical fenestration. It could also be considered as a model for complete fenestrated endovascular aortic repair (Kitagawa et al. 2013), or a 'post-operative' condition, in which the FL has been obliterated by TEVAR treatment and the true lumen diameter restored to healthy dimensions. Karmonik et al. (2011b) used inlet flow waveforms from MRI data, but applied constant pressure BCs at the branches and reported that their preliminary results correlated with clinical experience. Chen et al. (2013b) presented a number of treatment scenarios, including covering all entry tears, although they retained the TL diameter from the original scans (rather than expanding the TL as in Karmonik et al. (2011b)). They predicted that this would be the most appropriate treatment.

In this chapter, the virtual stenting approach was developed further by using the dynamic BCs that were deployed in the previous chapter. In addition to analysing the flow and WSS characteristics, the efficacy of the treatment options was considered by calculating the kinetic energy loss across the cardiac cycle for each case. In addition to a pre-operative model, two surgical solutions were simulated in the study. The 'single stent' case, in which the stent covers only the entry tear (standard TEVAR) and complete thrombosis is not achieved, and the 'double-stent' case in which both the entry and re-entry tears are fully occluded, and/or complete thrombosis and thus remodelling are achieved. This latter case models a stent extending down to the celiac artery (just distal to the re-entry tear), as recommended by Sayer et al. (2008); Kitagawa et al. (2013), although it could also be considered as a model of successful surgical fenestration (Karmonik et al. 2011b).

4.2 Methodology

The patient-specific geometry used in the previous chapter was also utilised in this chapter, and the same BCs were applied at each of the domains' boundary locations (Windkessel models) using FORTRAN. However, in order to make the results more

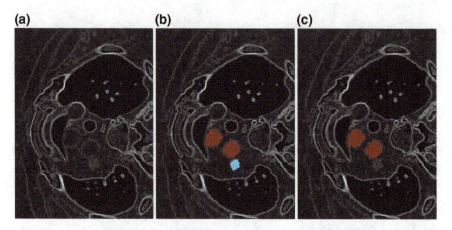

Fig. 4.1 CT scan slices showing **a** prior to the image segmentation, **b** showing selection of the regions of interest (*red*) and the aneurysm (*blue*), **c** removal of the aneurysm

general, the aneurysm that was located by the LS branch was virtually removed using ScanIP. The removal of the aneurysm can be carried out without affecting the rest of the geometry, and was implemented as follows: the stacks of images that included the aneurysm were detected and the removal was initiated at the first image in which the aneurysm was visible (Fig. 4.1a) going from the aortic arch superiorly (towards the head). Figure 4.1b shows two different masks, illustrating the regions that were selected (segmentation techniques were explained in Chap. 2). The red-coloured mask corresponds to the base of the BT and LS branch. There are only two branches originating from the aortic arch (shown by the red mask), as the patient had a bovine arch. The other mask (blue) defines the regions of the aneurysm. Two different masks were generated in order to enable the elimination of the blue mask (aneurysm) from the red mask (vessels). Figure 4.1c shows the image with the aneurysm removed.

Moving superiorly through the image stack, the area of the aneurysm increased and then subsequently decreased to zero. Some of the slices containing the aneurysm did not have well defined borders between the aneurysm and the LS branch, therefore virtual creation of the LS branch for these regions was prescribed. To virtually re-create this branch, the last slice detecting this branch was selected and tapering was applied, i.e. at each sequential image going superiorly, an area reduction of 0.15 pixels was applied until the borders between the aneurysm and LS branch became visible. In addition, a similar process was carried out to remove the bovine aortic arch, in order to obtain a more general model.

The resulting geometry is shown in Fig. 4.2 and represents the pre-operative case. Entry and re-entry tears are highlighted in the pre-operative case (Fig. 4.2), with average diameters of 23.8 and 21.8 mm respectively; the length of the domain is 344 mm with the FL having a length of 317 mm. The branch outlets were the same for all cases (pre- and post-treatment) and had the following equivalent diameters ($D \approx \sqrt{\frac{4A}{\pi}}$),

4.2 Methodology

Fig. 4.2 Geometry of a patient-specific dissected aorta after removing the aneurysm. Inset; *upper panel* shows CT slice at the *dashed line*; *lower panel* shows CT slice, highlighting the ascending aorta and TL in *blue* and the FL in *green*

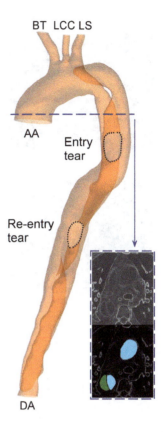

ascending aorta: 33.2 mm; brachiocephalic trunk: 11.1 mm; left common carotid: 6.8 mm; left subclavian: 7.5 mm and distal abdominal: 17.1 mm. The diameter of the coarctation in the pre-operative case was 10.2 mm. For the post-operative cases the diameter at the same location was increased to 33.8 mm. The inset in Fig. 4.2 shows a sample CT image slice at the location indicated by the dashed blue line. The regions selected in ScanIP are indicated, with green representing the FL and blue representing the ascending aorta and TL.

Three separate computational 3D domains (pre-operative, single and double stent) were created. To simulate the stented domain (single stent-graft case), the entry tear was virtually sealed and the intima flap regions were removed. The surgical procedure of sealing the entry tear stops blood entering the FL, which becomes thrombosed, leading to aortic remodelling, resulting in increased the TL cross-sectional area and decreased FL dimensions. However, complete FL thrombosis is not always achieved when only blocking the entry tear (Scali et al. 2013a; Mani et al. 2012). The 'single-stent case' represents successful thrombosis of the proximal FL, but not the distal FL, and thus the dissected region remains after the second tear. Figure 4.3 shows the stented geometry; the stented region starts proximal to the entry tear and ends just proximal to the re-entry tear.

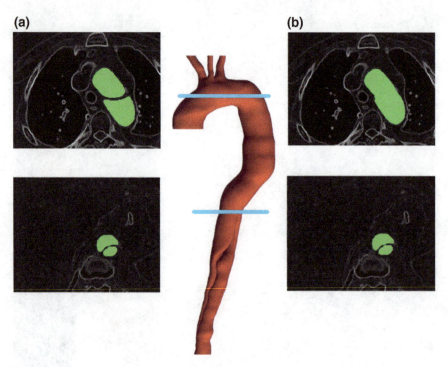

Fig. 4.3 Patient-specific single-stented reconstructed geometry. **a** Pre-virtual stenting, **b** post-virtual stenting. *Blue lines* correspond to the location of slices selected

As it is often preferable to stent the entire thoracic aorta (Kölbel et al. 2014), a second surgical scenario was also investigated, simulating the double stent-graft that additionally occludes the re-entry tear and is referred to as the double-stent case in this study (Fig. 4.4). Figure 4.4a, b show the pre- and post-stenting on the left and right respectively. In the double-stent case, the distal FL is eliminated by a virtual stent (filling the gap visible in the green mask; Fig. 4.4a). All three cases are shown in Fig. 4.5.

Each of the three 3D geometries were imported into ANSYS ICEM meshing software and were discretised into approximately 250,000 tetrahedral mesh elements with 7 prismatic layers at the wall for each case, similarly to Chap. 3. In order to ensure consistency between the three cases, the same mesh settings were used for each geometry. A mesh sensitivity analysis was carried out for the three geometries and is discussed in Sect. 4.5 and Appendix B. As a compromise between accuracy and computational efficiency, the medium mesh was used for the main analysis.

4.2 Methodology

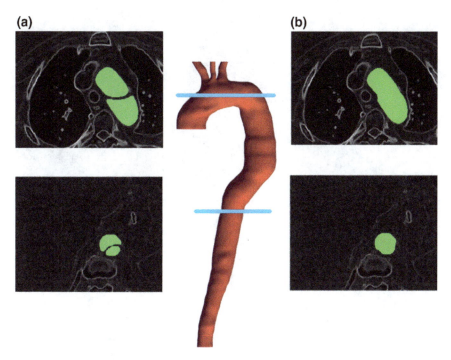

Fig. 4.4 Patient-specific double stented reconstructed geometry. **a** Pre-virtual stenting, **b** post-virtual stenting. *Blue lines* correspond to the location of slices selected

4.2.1 Boundary Conditions

At the inlet of the three models, a flow wave was prescribed, which was taken from a study performed by Karmonik et al. (2008), as patient-specific flow data was not available for the present study. The waveform is shown in Fig. 4.6 for the ascending aorta. The patient had a heart rate of 70 beats/min and thus the cardiac cycle time was adjusted accordingly (0.857 s for one cardiac cycle, $f = 1.17$ Hz) giving a cardiac output of 4.25 l/min.

Blood was considered to be an incompressible, Newtonian fluid with a density of $\rho = 1056$ kg/m^3 and dynamic viscosity of $\mu = 3.5$ mPa s. For this study, a laminar model was utilised. Stalder et al. (2011) determined a critical Reynolds number for the aorta given by:

$$Re_c = 169\alpha^{0.83} St^{-0.27} \tag{4.1}$$

where $\alpha = 0.5D\sqrt{2\pi f \rho \mu^{-1}}$ and $St = 0.5fD(U_p - U_m)^{-1}$ are the Womersley and Strouhal numbers respectively. U_p and U_m are the peak and mean velocities, which were 0.42 and 0.08 m/s respectively. The Womersley number in the ascending aorta was 24.7, the Strouhal number was 0.058, the mean and peak Reynolds numbers

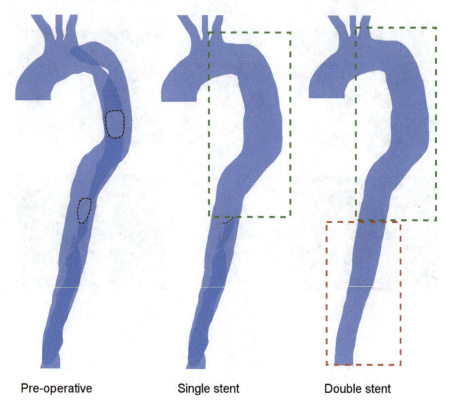

Fig. 4.5 Pre-operative, single-stent (entry tear occluded) and double-stent (both tears occluded) geometries. The *dashed boxes* indicate the entry tear and re-entry tear in the pre-operative case

were approximately 800 and 4100. From these values, the critical Reynolds number was calculated to be 5200. Based on these values, the inlet flow is expected to be subcritical throughout the cycle. Combined with the small influence of turbulence on the results observed in Sect. 3.7, this implies that a laminar model is reasonable for this analysis.

As in the previous chapter, all three models were coupled at their outlet boundaries to three element Windkessel models. R_1, R_2 and C parameters at each boundary were identical to those used in the previous chapter for all three models. These parameters represent the downstream vasculature, and thus implementation of the same boundary conditions for all three cases enabled the investigation of haemodynamic changes due to stenting, such as the energy loss across a cardiac cycle.

In the interest of computational efficiency and in order to aid the direct comparison between the different cases considered herein, rigid walls were used in the simulations. Patients suffering from AD have been found to exhibit decreased wall distensibility (Tse et al. 2011; Pyeritz 2000), although some wall motion is still expected. Furthermore, as the stent-graft itself has different mechanical properties to

4.2 Methodology

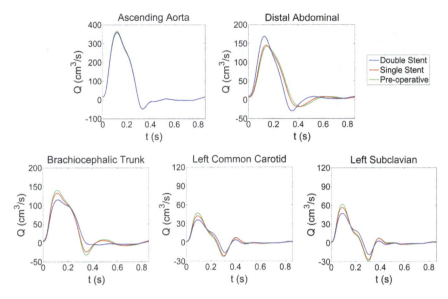

Fig. 4.6 Flow waveforms at the inlet and outlets of all three cases for one cardiac cycle

the vessel wall, accurate modelling of the solid mechanics is non-trivial and hence is beyond the scope of these preliminary analyses. The influence of FSI techniques on CFD simulations of AD will be considered in Chap. 5.

The time-step used for all three simulations was 5 ms. The results were taken once the simulation had reached a periodic state, which took two cycles for all cases studied (after appropriate initialisation). Therefore, the results of the third cycle are investigated here.

4.3 Results

4.3.1 Velocity and Flow Rates

Figure 4.6 shows the flow waveforms at the boundaries of the 3D domain. In the branches, the maxima and minima of the waves are reduced in magnitude by the successive stenting operations. In the DA, however, these increased significantly in the double-stent case, but not in the single-stent case. This change indicates a reduction in the flow resistance in the DA for the double-stent, but not the single-stent case.

Figure 4.7 shows velocity magnitude contours for all three cases at the time of peak systolic pressure. The velocity is similar in the branches for all three models. In the pre-operative case, the velocity magnitude is high in the coarctation region

Fig. 4.7 Velocity magnitude contours at peak systolic flow for all three cases

(distal to the arch) and the distal TL, whereas small values can be seen in the proximal and distal FL. The single-stenting operation removed the coarctation and proximal FL, and hence the velocity in the proximal descending aorta was more uniformly distributed than in the pre-operative case. Occluding the re-entry tear (double-stent case), resulted in a significant decrease in the velocity along the abdominal aorta.

The streamlines for all three cases at peak systolic pressure are shown in Fig. 4.8. In the pre-operative case, the streamlines are very uniform along the aortic arch and immediately downstream of the arch, prior to the first tear (entry tear). In both the proximal and distal FL, complex vortical structures can be seen. In the superior region of the proximal FL, no streamlines are visible, as negligible flow from the domain inlet and outlets, from where the streamlines are calculated, enters this region (Fig. 4.8, pre-operative case). The streamlines are fairly uniform in between the two tears in both TL and FL. In the single-stent case, the insertion of the stent extends the region of the uniform flow through the domain (due to removal of the proximal FL and thus expansion of the coarctation), however the flow in the distal FL remains disordered. The double-stenting operation (Fig. 4.8) smoothens the flow throughout the domain and no vortical structures can be observed at peak systole.

4.3 Results 111

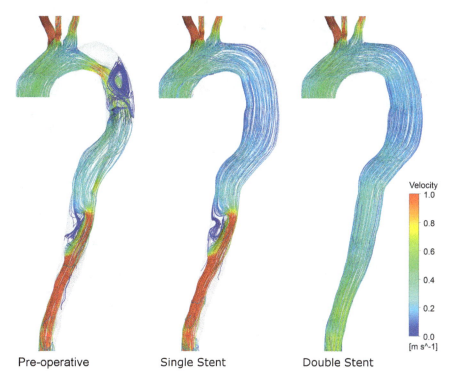

Fig. 4.8 Streamlines at peak systolic flow for all three cases

4.3.2 Pressure

Figure 4.9 compares the distributions of pressure at peak systole. In the pre-operative case, pressure is greatest along the aortic arch and decreases at the branches. A large pressure drop can be observed in the downstream TL. In the FL, the pressure was fairly uniform. Downstream of the re-entry tear, the pressure gradient between the two lumina increased with the distance from the tear.

By occluding the entry tear (single-stent case) and hence also removing the coarctation, the pressure gradient along the distal arch region is greatly reduced, resulting in a lower mean pressure in the arch. However, the dissected region of the thoracic aorta still results in a relatively large pressure gradient due to the significantly reduced cross sectional area. Occluding the second tear (double-stent) appears to remedy this problem, and as a result, the pressure is uniformly distributed throughout the aorta (note that the same pressure scale has been used and so the colour varies little throughout the domain, even though there is a net pressure gradient).

The pressure waveforms at the domain boundaries are shown in Fig. 4.10. At the AA, the pulse pressure was decreased by both stenting operations (approximately 2.5 mmHg for each occlusion step). The pressure wave displays the bisferiens

Fig. 4.9 Pressure contours at peak systole for all three cases

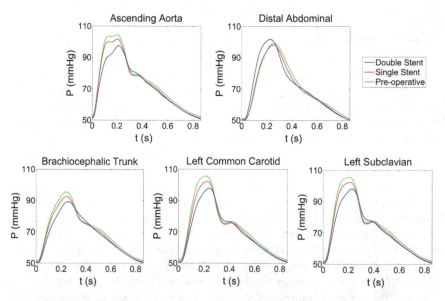

Fig. 4.10 Comparison of pressure waves at the boundaries for pre- and post-operative cases

characteristic in both the pre-operative and single-stent cases; however the double-stenting procedure seems to attenuate this. This supports the idea that that the observed bisferiens-like pressure wave observed in the pre-treated case may occur due to the abdominal reduction in the TL cross-sectional area. A similar reduction in peak pressure with the virtual treatments can be observed in the pressure waveforms for the LCC, BT and LS outlets. For the DA, the insertion of the single stent-graft resulted in a lower pulse pressure and a small phase shift (relative to the pre-operative case), whereas occluding both tears increased the pulse pressure (higher maximum pressure) and its phase shift in relation to the pre-operative waveform.

4.3.3 Kinetic Energy

In order to analyse the reduction in flow resistance caused by the stenting operations, the kinetic energy loss (assuming negligible gravitational effect) in the system was analysed, similarly to the approach used by Lee et al. (2013). The rate of kinetic energy loss ($\dot{K}E$) at each boundary i at each time instance is given by

$$\dot{K}E_i(t) = Q_i(t)\left(P_i(t) + \frac{1}{2}\rho \bar{V}_i^2(t)\right) \quad (4.2)$$

The total KE loss across the cardiac cycle is given by integrating the difference between the KE at the inlet and outlets:

$$\Delta KE = \int_0^T \left[\dot{K}E_{AA}(t) - \dot{K}E_{DA}(t) - \dot{K}E_{BT}(t) - \dot{K}E_{LCC}(t) - \dot{K}E_{LS}(t)\right] dt \quad (4.3)$$

For the pre-operative, single- and double-stent cases, the energy loss per cycle was calculated to be 55.2, 54.4 and 32.9 mJ respectively. Thus, the single- and double-stent operations respectively reduced the energy loss by 1.5% and 40.4%.

4.3.4 Wall Shear Stress

As discussed in the previous chapter, in AD patients, elevated values of WSS are correlated with tear enlargement, additional tear generation and aneurysm formation (Mohr-Kahaly et al. 1989). In aneurysms, low WSS is associated with increased risk of aneurysm expansion or rupture by disturbing the endothelial functions (Shojima 2004), and collocated regions of low TAWSS and high OSI are more prone to rupture (Xiang et al. 2010). Although it is not clear how dissected regions respond to wall shear characteristics, it is reasonable to expect that they might behave similarly to aneurysms, given that ex vivo studies have shown that the false lumen becomes completely endothelialised (Roberts 1981). TAWSS and OSI are thus important

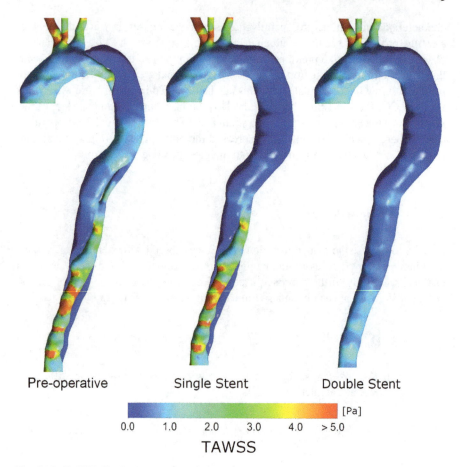

Fig. 4.11 TAWSS distribution for all three cases

parameters in AD; their distributions are shown in Figs. 4.11 and 4.13 for both pre- and post-operative cases.

In the pre-operative case of Fig. 4.11, high values of TAWSS can be observed at the branches, around the entry tear (left posterior view not shown for brevity), distal to the re-entry tear and along the distal TL. In the proximal and distal FL, the TAWSS is very low. For the single stent-graft case, the TAWSS characteristics are similar to the pre-operative case along the aortic arch and the three branches. However, the TAWSS is reduced in the distal arch and in the TL. In the double-stent case, the TAWSS distribution in the proximal aortic arch and the branches is similar to those observed in the other cases; however TAWSS values are greatly reduced in the descending aorta.

Figure 4.12 provides histograms of the TAWSS values for the three geometries. The histograms are calculated based on the number of surface elements with a given TAWSS value and are normalised to give a percentage of the total number of elements.

4.3 Results

Fig. 4.12 Histograms of TAWSS. *Dashed vertical lines* show 5th and 95th percentiles, *solid vertical line* shows median. **a** Pre-operative, **b** single-stent, **c** double-stent

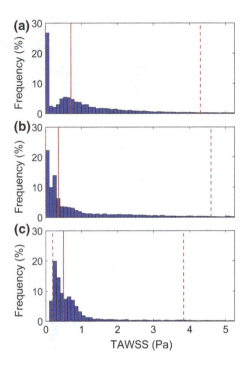

Note that although the mesh elements vary in size, the counts provide an indication of the distribution of the parameters averaged over the whole surface. Dashed red lines show the 5th and 95th percentiles, while the solid red line shows the median (50th percentile). For the pre-operative case (Fig. 4.12a), almost 25% of the surface elements exhibit very low TAWSS (<0.1 Pa), which are located in the proximal and distal FL. Additionally, a large number of elements with high TAWSS are present (38% > 1 Pa). For the single-stent case (Fig. 4.12b), the high stress values in the coarctation are absent, and thus the median is significantly decreased. However, a large number of elements still exhibit a very low TAWSS (predominantly in the distal FL). Here, 30% of the elements have a TAWSS > 1 Pa. For the double-stent case, the extremely low WSS regions are no longer apparent, and hence no elements are observed with TAWSS <0.1 Pa. However, as a result, the median value is increased relative to the single-stent case. Furthermore, the regions of elevated WSS are decreased, with only 19% having WSS >1 Pa. Note that neither minimising or maximising WSS is ideal (in terms of prognosis), and extremes are likely to incur greater risk (Meng et al. 2014).

The OSI distributions in Fig. 4.13 indicate scattered regions of high OSI throughout the domain for the pre-operative case, for example along the aortic arch and branches. In the FL, the OSI is elevated around the entry tear, in between the two tears and in the distal and proximal (to a lesser extent) FL. Small regions of high OSI can be seen along the distal TL. The OSI behaviour observed along the distal TL in

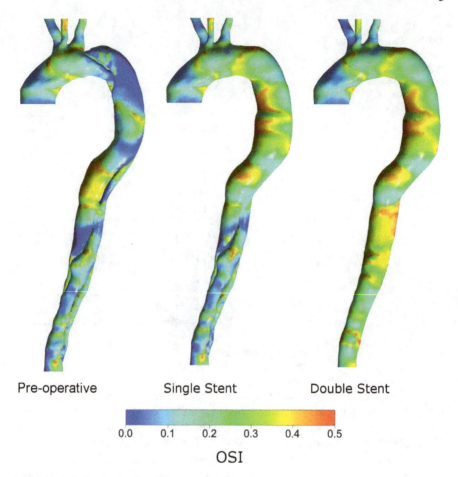

Fig. 4.13 OSI distribution for all three cases (*right*)

the single-stent case is similar to the one observed in the pre-operative case. In the aortic arch and proximal descending aorta, the OSI is similar for both stent cases. However, the addition of the second stent increased OSI in the thoracic aorta.

Figure 4.14 shows histograms of OSI with the 5th and 95th percentiles indicated by vertical lines. It can be seen that the pre-operative case has a strongly skewed distribution, with a relatively large number of surface elements exhibiting low OSI values. The single- and double-stent operations shifted the median to higher values, and made the distribution more symmetric. As with TAWSS, extreme values of OSI may indicate a pathological environment for endothelial cells, implying that the double-stent operation might reduce risk of future complications.

Thus, after the virtual double-stent operation to remove the dissection, the OSI increased on average (Fig. 4.14). Note that the TAWSS at a given location is the average magnitude of the wall shear stress $\left(\frac{1}{T}\int_0^T |\tau(t)|\, dt\right)$, whereas the OSI is

4.3 Results

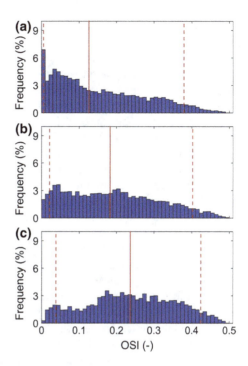

Fig. 4.14 Histograms of OSI. Dashed vertical lines show 5th and 95th percentiles, *solid vertical line* shows median. **a** Pre-operative, **b** single-stent, **c** double-stent

proportional to the ratio of the magnitude of the average wall shear stress $\left(\left|\frac{1}{T}\int_0^T \tau(t)dt\right|\right)$ to the TAWSS. If the flow was unidirectional, the two terms would be equal and the OSI would be 0 (see Eq. 3.5). If the flow was perfectly sinusoidal, the average wall shear stress vector would have a magnitude of zero and the OSI would be maximal (0.5). When the flow acts in various directions throughout the cycle, as observed in the diseased pre-operative case, the oscillations do not cancel out, and thus the average WSS vector has significant components in all directions, resulting in a low OSI. In the double-stent case, the simpler geometry means that the flow wave is closer to the ideal sinusoid and thus the average WSS vector has a lower magnitude and the OSI is greater.

4.4 Discussion

In the present study, 3D simulations of the flow through an AD in pre- and post-operative states involving two virtual stent-graft operations were carried out. The results of the study showed significant haemodynamic differences between the two stenting operations. Studies have shown that the blood flow through the FL is one of the main risk factors for its enlargement (Grabenwoger et al. 2012; Patel et al. 2009;

Sueyoshi 2004). However, by occluding the entry tear this potential problem can be addressed (Weigang et al. 2008b).

From the results, it was observed that both the maxima and minima of the flow wave in the descending aorta were increased by the double-stent procedure. This indicates that the resistance of the thoracic aorta (DA) was decreased, which is important when treating patients with lower limb malperfusion syndrome. It should be emphasised that while the first single stent-graft reduced the pressure in the aortic arch, it did not significantly affect the flow in the DA.

The simulations indicated high pressure gradients around the aortic arch and between the TL and FL in the pre-operative case. However, occluding the entry tear (single stent-graft) decreased the pressure gradient along the dissected aorta in the stented region, but did not significantly affect the thoracic aorta, wherein a dissection was still present. In the double-stent case, the pressure gradient was greatly reduced. The single stent-graft procedure lowered the pulse pressure by 2.5 mmHg. The double stent-graft reduced the pulse pressure by a further 2.5 mmHg. This is not surprising as TEVAR has been used to manage hypertension in patients with AD (Khoynezhad et al. 2009). The importance of this reduction can be better understood by considering the energy loss in the system; the analysis showed that the double stent-graft operation reduced the energy loss by 40%, compared to only 1.5% for the single stent-graft. From a fluid dynamics perspective, the high flow resistance generated by the partial occlusion of the DA required significantly more energy. As a corollary, these results predict that the double-stent procedure would significantly reduce the load on the heart, whilst the single-stent procedure would have a negligible effect. Additionally, the flow in the stented geometries was more uniform, altering the WSS distributions.

In aneurysms, low WSS is associated with increased risk of aneurysm expansion or rupture by disturbing the endothelial functions (Xiang et al. 2010). Additionally, studies have shown that high values of OSI and low values of TAWSS indicate regions more prone to rupture (Roberts 1981). Figure 4.15 shows the distribution of the product TAWSS × (0.5-OSI), where low values yield regions where both high OSI and low WSS are located. It can be seen that the pre-operative and single-stent cases have the same number of elements in the lowest value bin, which could be associated with high risk of rupture (Roberts 1981). This indicates that the region of greatest risk is the distal FL, which is not affected by the single-stent operation. In the double-stent case, the number of elements in the high risk area is less than 1%. It should be noted that the relative residence time (RRT), an index which is used to reflect the residence time of blood near the vessel wall, is equal to half the reciprocal of TAWSS × (0.5-OSI) (Xiang et al. 2010).

By occluding both tears, the TAWSS distribution became more uniform, the average OSI was almost doubled and the distribution of OSI became more uniform, indicating a healthier mechanical environment for the endothelial cells lining the aorta (Hoi et al. 2011).

The majority of the limitations of this chapter are similar to those described in Sect. 3.4.1, and so are not repeated here. In addition, the influence of the modification of the geometry (removal of the aneurysm and bovine arch) on the viability of the Windkessel parameters should be considered. Given that the coupled

4.4 Discussion

Fig. 4.15 Histograms of the product TAWSS × (0.5-OSI). **a** Pre-operative, **b** single-stent, **c** double-stent

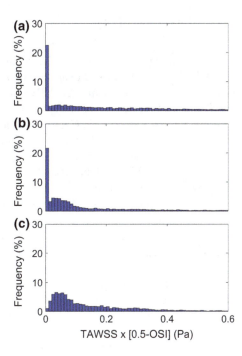

Windkessel models represent the downstream vasculature, modification of the 3D domain should not significantly affect the validity of the RCR parameters that were defined in Chap. 3. To clarify, the Windkessel models output a pressure, as a function of the flow rate and RCR parameters. In the paradigm of the modelling approach, the RCR parameters are purely dependent on the downstream vasculature, even though the estimated parameters are defined based on the interaction with the 3D domain.

Furthermore, although long-term remodelling of the downstream vasculature may also alter the optimal RCR values as the patient recovered from the virtual surgery, the present results can be viewed as the short term response.

The difference in the shear-stresses on the surface of the stent between the difference models was not significant. However, the pressure forces contribute significantly more to the total force acting on the stent than the shear stress (Figueroa et al. 2009) (several orders of magnitude). By reducing the pulse pressure, the double stent procedure reduced the pressure (normal stress) on the stent, although only by approximately 10%. Thus there is no expected difference in the likelihood of stent migration between the two models.

In order to fully validate this approach, longitudinal clinical data, in which pre- and post-operation clinical data was available for a number of patients, would be necessary. Blind studies could be carried out in which the pre-operation geometry and data, along with the desired treatment geometry, are provided and the post-operation data is held back. If a model was able to predict the observed changes, then its ability as a predictive tool would be confirmed.

4.5 Mesh Sensitivity

As mentioned previously, in order to assess the sensitivity of the results to the number of mesh elements, two additional meshed domains were also examined for each geometry; a coarse mesh with ≈80,000 elements and a fine mesh with approximately ≈600,000 mesh elements. The coarse mesh utilised 4 prism layers, whilst the medium and fine meshes used 7, as in Chap. 3. The mesh size settings were defined for the pre-operative case to achieve the desired number of mesh elements and the same settings were then applied to the virtually stented geometries.

Appendix B contains a detailed analysis of the mesh sensitivity effects on the flow and pressure waves at the boundaries and the distributions of velocity, pressure and WSS indices. Table 4.1 lists the differences in flow and pressure estimates between the three meshes. Pressure values are given as a proportion of the pulse pressure and flow rates are presented as a percentage of the maximum value in each branch. The values shown represent the maximum differences observed across the inlet and all four outlets.

It can be seen that the pre-operative case is more sensitive to changes in the mesh than the single- and double-stented cases. The largest difference is observed for the pressure estimated in the pre-operative case. For the pre-operative and double-stent cases, the difference between the medium and fine meshes is less than for the coarse and medium meshes. For the single-stent case the difference between the medium to fine mesh is slightly larger than for the coarse to medium mesh, but the percentage difference is very small.

The spatial distributions of TAWSS and OSI values obtained using different meshes are given in Appendix B. Here, histograms comparable to Figs. 4.12 and 4.14 are shown to illustrate the mesh dependence of the WSS. Each column in Fig. 4.16 shows a treatment case, and each row shows the results of the three meshes. The blue histograms of the medium mesh WSS estimates show the same data as Fig. 4.12. It can be seen that overall, the shape of the distributions does not change between the three meshes studied. This suggests that the histograms represent the distributions of the TAWSS appropriately (i.e. the different element sizes do not significantly affect the analysis).

Table 4.1 Maximum differences between the pressure and flow estimates at the boundaries obtained with different meshes

		Coarse—Medium	Medium—Fine
Flow rate (%)	Pre-operative	1.6	1.2
	Single-stent	0.9	1.5
	Double-stent	0.3	0.2
Presure (%)	Pre-operative	5.4	2.4
	Single-stent	2.8	3.0
	Double-stent	0.9	0.2

4.5 Mesh Sensitivity

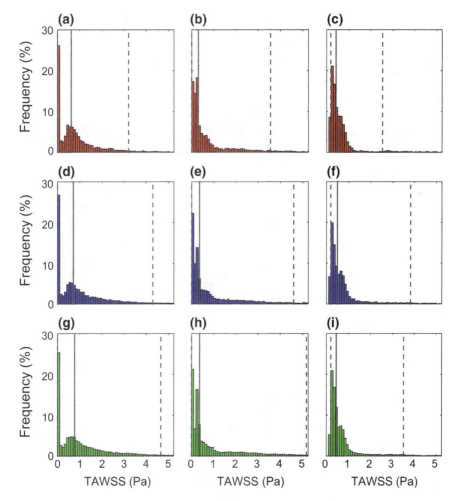

Fig. 4.16 Histograms of TAWSS for the mesh sensitivity analysis. *Dashed vertical lines* show 5th and 95th percentiles, *solid vertical line* shows median. *Red* coarse mesh, *Blue* medium mesh, *Green* fine mesh. **a, d, g** Pre-operative, **b, e, h** single-stent, **c, f, i** double-stent

For the pre-operative case, the 95th percentile (indicating regions of high WSS) increases with sequential mesh refinement (from Fig. 4.16a to d to g). The same is observed for the single-stent case, (from Fig. 4.16b to e to h). For the double-stent case there is less variability between the medium and fine meshes (Fig. 4.16f and i) and the latter shows a marginally lower 95th percentile. As the double-stent geometry is considerably less tortuous than the single-stent or pre-operative aortae, it could be expected that the results would be less sensitive to mesh refinement, as supported by this data.

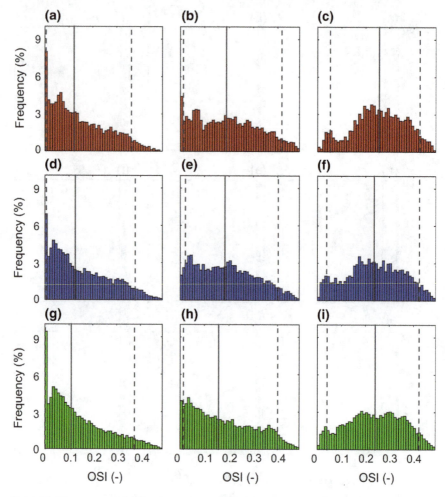

Fig. 4.17 Histograms of OSI for the mesh sensitivity analysis. *Dashed vertical lines* show 5th and 95th percentiles, *solid vertical line* shows median. *Red* coarse mesh, *Blue* medium mesh, *Green* fine mesh. **a, d, g** Pre-operative, **b, e, h** single-stent, **c, f, i** double-stent

Mesh refinement did not significantly affect the statistical distributions of OSI, as can be seen in Fig. 4.17. For all three cases, the 5th, 50th and 95th percentiles are essentially the same for all three meshes. However, some local variations were observed (see Appendix B).

In conclusion, all three meshes produced comparable results, and all deviations observed were relatively minor. Comparisons between the coarse and medium meshes showed that there were notable differences in the localised WSS

4.5 Mesh Sensitivity

predictions indicating that areas of high WSS might be missed by the coarse mesh (see Appendix B). Differences between the estimated pressure and flow waveforms tended to be larger between the coarse and medium meshes, than between the medium and fine meshes.

Simulation time is a key consideration if such computational approaches are to be translated into a clinical setting. Considering for example the single-stent case, the simulation with the medium mesh took just over twice as long as with the coarse mesh, and the simulation with the fine mesh required more than twice the time taken for the simulation with the medium mesh. Thus, for the present study, the medium mesh was selected for the analysis, as it required a reasonable simulation time, while producing results that were not significantly different to those of the fine mesh.

Finally, it should be noted that the dependency of the results on the mesh refinement decreased for the stented geometries, as the tortuosity decreased. Hence, in a more time-critical clinical setting, if mesh sensitivity was analysed, it could potentially be carried out only on the pre-operative geometry in order to increase efficiency.

4.6 Conclusion

This chapter compared and evaluated two interventional strategies for a patient with type-B AD. CFD simulations coupled with Windkessel models were performed. The scope of the present study was to compare the haemodynamic parameters between pre-operative, single and double-stent cases and to provide further development of personalised workflows. In particular, the data presented suggests that the double-stent graft option, although potentially more risky, would be significantly more effective than occlusion of the first tear only for this patient. Although the double stent-graft procedure would have increased risk and complexity associated with it, the present results showed a 40% reduction in flow resistance, compared to just 1.5% for the single-stent. Furthermore, the double stent-graft procedure further reduced the pulse pressure and significantly attenuated regions of elevated WSS in the descending aorta. The method could be easily transferred to the clinic to aid clinicians with the decision making process via evaluating and comparing alternative intervention strategies. In order to introduce simulations to the clinic, a number of improvements to modelling approaches and further validation with larger numbers of patients are still required.

Both the present and previous chapters made the assumption that the vessel walls were rigid, as is common in simulations of AD (Chen et al. 2013b; Cheng et al. 2010; Tse et al. 2011). In the next chapter, the influence of wall motion on the analysis of AD is analysed using FSI simulations.

References

Alimohammadi, M., Bhattacharya-Ghosh, B., Seshadri, S., Penrose, J., Agu, O., Balabani, S., et al. (2014b). Evaluation of the hemodynamic effectiveness of aortic dissection treatments via virtual stenting. *International Journal of Artificial Organs, 37*(10), 753–762.

Chen, D., Müller-Eschner, M., Rengier, F., Kotelis, D., Böckler, D., Ventikos, Y., et al. (2013c). A preliminary study of fast virtual stent-graft deployment: application to stanford type B aortic dissection. *International Journal of Advanced Robotic Systems, 10*, 154.

Chen, D., Müller-Eschner, M., Kotelis, D., Böckler, D., Ventikos, Y., & von Tengg-Kobligk, H. (2013b). A longitudinal study of Type-B aortic dissection and endovascular repair scenarios: computational analyses. *Medical Engineering and Physics, 35*(9), 1321–1330.

Cheng, Z., Tan, F. P. P., Riga, C. V., Bicknell, C. D., Hamady, M. S., Gibbs, R. G. J., et al. (2010). Analysis of flow patterns in a patient-specific aortic dissection model. *Journal of Biomechanical Engineering, 132*(5), 051007.

Criado. (2011). Aortic dissection: A 250-year perspective. *Texas Heart Institute Journal, 38*(6), 694–700.

Fattori, R., Tsai, T. T., Myrmel, T., Evangelista, A., Cooper, J. V., Trimarchi, S., et al. (2008b). Complicated acute type b dissection: is surgery still the best option? *JACC: Cardiovascular Interventions, 1*(4), 395–402.

Fattori, R., Cao, P., De Rango, P., Czerny, M., Evangelista, A., Nienaber, C., et al. (2013). Interdisciplinary expert consensus document on management of type B aortic dissection. *Journal of the American College of Cardiology, 61*(16), 1661–1678.

Figueroa, C. A., Taylor, C. A., Chiou, A. J., Yeh, V., & Zarins, C. K. (2009). Magnitude and direction of pulsatile displacement forces acting on thoracic aortic endografts. *Journal of Endovascular Therapy, 16*(3), 350–358.

Gopaldas, R. R., Huh, J., Dao, T. K., LeMaire, S. A., Chu, D., Bakaeen, F. G., et al. (2010). Superior nationwide outcomes of endovascular versus open repair for isolated descending thoracic aortic aneurysm in 11,669 patients. *The Journal of Thoracic and Cardiovascular Surgery, 140*(5), 1001–1010.

Grabenwoger, M., Alfonso, F., Bachet, J., Bonser, R., Czerny, M., Eggebrecht, H., et al. (2012). Thoracic Endovascular Aortic Repair (TEVAR) for the treatment of aortic diseases: a position statement from the European Association for Cardio-Thoracic Surgery (EACTS) and the European Society of Cardiology (ESC), in collaboration with the European Association of Percutaneous Cardiovascular interventions (EAPCI). *European Heart Journal, 33*, 1558–1563.

Hagan, P. G., Nienaber, C. A., Isselbacher, E. M., Bruckman, D., Karavite, D. J., Russman, P. L., et al. (2000). "The international registry of acute aortic dissection (IRAD). *JAMA: The Journal of the American Medical Association, 283*(7), 897–903.

Hofferberth, S. C., Nixon, I. K., & Mossop, P. J. (2012). Aortic false lumen thrombosis induction by embolotherapy (AFTER) following endovascular repair of aortic dissection. *Journal of Endovascular Therapy, 19*(4), 538–545.

Hoi, Y., Zhou, Y.-Q., Zhang, X., Henkelman, R. M., & Steinman, D. A. (2011). Correlation between local hemodynamics and lesion distribution in a novel aortic regurgitation murine model of atherosclerosis. *Annals of Biomedical Engineering, 39*(5), 1414–1422.

Karmonik, C., Bismuth, J. X., Davies, M. G., & Lumsden, A. B. (2008). Computational hemodynamics in the human aorta: a computational fluid dynamics study of three cases with patient-specific geometries and inflow rates. *Technology and Health Care, 16*(5), 343–354.

Karmonik, C., Bismuth, J., Shah, D. J., Davies, M. G., Purdy, D., & Lumsden, A. B. (2011b). Computational Study of haemodynamic effects of entry- and exit-tear coverage in a DeBakey type III Aortic Dissection: Technical Report. *European Journal of Vascular and Endovascular Surgery, 42*(2), 172–177.

Khoynezhad, A., Donayre, C. E., Omari, B. O., Kopchok, G. E., Walot, I., & White, R. A. (2009). Midterm results of endovascular treatment of complicated acute type B aortic dissection. *The Journal of Thoracic and Cardiovascular Surgery, 138*(3), 625–631.

References

Kitagawa, A., Greenberg, R. K., Eagleton, M. J., Mastracci, T. M., & Roselli, E. E. (2013). Fenestrated and branched endovascular aortic repair for chronic type B aortic dissection with thoracoabdominal aneurysms. *Journal of Vascular Surgery*, *58*(3), 625–634.

Kölbel, T., Tsilimparis, N., Wipper, S., Larena-Avellaneda, A., Diener, H., Carpenter, S. W., et al. (2014). TEVAR for chronic aortic dissection—is covering the primary entry tear enough? *The Journal of Cardiovascular Surgery*, *55*(4), 519–527.

Lee, N., Taylor, M. D., Hor, K. N., & Banerjee, R. K. (2013). Non-invasive evaluation of energy loss in the pulmonary arteries using 4D phase contrast MR measurement: a proof of concept. *Biomedical Engineering Online*, *12*(1), 93.

Loubert, M. C., van der Hulst, V. P. M., De Vries, C., Bloemendaal, K., & Vahl, A. C. (2003). How to exclude the dilated false lumen in patients after a type B aortic dissection? The cork in the bottleneck. *Journal of Endovascular Therapy*, *10*(2), 244–248.

Mani, K., Clough, R. E., Lyons, O. T. A., Bell, R. E., Carrell, T. W., Zayed, H. A., et al. (2012). Predictors of outcome after endovascular repair for chronic type B dissection. *European Journal of Vascular and Endovascular Surgery*, *43*(4), 386–391.

McPhee, J. T., Asham, E. H., Rohrer, M. J., Singh, M. J., Wong, G., Vorhies, R. W., et al. (2007). The midterm results of stent graft treatment of thoracic aortic injuries. *The Journal of Surgical Research*, *138*(2), 181–188.

Meng, H., Tutino, V. M., Xiang, J., & Siddiqui, A. (2014). High WSS or low WSS? Complex interactions of hemodynamics with intracranial aneurysm initiation, growth, and rupture: toward a unifying hypothesis. *AJNR. American Journal of Neuroradiology*, *35*(7), 1254–1262.

Mohr-Kahaly, S., Erbel, R., Rennollet, H., Wittlich, N., Drexler, M., Oelert, H., et al. (1989). Ambulatory follow-up of aortic dissection by transesophageal two- dimensional and color-coded Doppler echocardiography. *Circulation*, *80*(1), 24–33.

Nienaber, C., Fattori, R., Lund, G., Dieckmann, C., Wolf, W., von Kodolitsch, Y., et al. (1999). Non-surgical reconstruction of thoracic aortic dissection by stent-graft placement. *The New England Journal of Medicine*, *340*, 1539–1545.

Nienaber, C. A. (2011). Influence and critique of the INSTEAD trial (TEVAR Versus Medical Treatment for Uncomplicated Type B Aortic Dissection). *Seminars in Vascular Surgery*, *24*(3), 167–171.

Nordon, I. M., Hinchliffe, R. J., Loftus, I. M., Morgan, R. A., & Thompson, M. M. (2011). Management of acute aortic syndrome and chronic aortic dissection. *Cardiovascular and Interventional Radiology*, *34*(5), 890–902.

Patel, H. J., Williams, D. M., Meekov, M., Dasika, N. L., Upchurch, G. R, Jr., & Deeb, G. M. (2009). Long-term results of percutaneous management of malperfusion in acute type B aortic dissection: Implications for thoracic aortic endovascular repair. *The Journal of Thoracic and Cardiovascular Surgery*, *138*(2), 300–308.

Pyeritz, R. E. (2000). The Marfan Syndrome. *Annual Review of Medicine*, *51*, 481–510.

Roberts, W. (1981). Aortic dissection: Anatomy, consequences, and causes. *American Heart Journal*, *101*(2), 195–214.

Sayer, D., Bratby, M., Brooks, M., Loftus, I., Morgan, R., & Thompson, M. (2008). Aortic morphology following endovascular repair of acute and chronic type B aortic dissection: implications for management. *European Journal of Vascular and Endovascular Surgery*, *36*(5), 522–529.

Scali, S. T., Feezor, R. J., Chang, C. K., Stone, D. H., Hess, P. J., Martin, T. D., et al. (2013a), Efficacy of thoracic endovascular stent repair for chronic type B aortic dissection with aneurysmal degeneration. *Journal of Vascular Surgery*, *58*(1), 10–7.e1.

Scali, S. T., Waterman, A., Feezor, R. J., Martin, T. D., Hess, P. J., Huber, T. S., et al. (2013b). Treatment of acute visceral aortic pathology with fenestrated/branched endovascular repair in high-surgical-risk patients. *Journal of Vascular Surgery*, *58*(1), 56–65.e1.

Shojima, M. (2004). Magnitude and role of wall shear stress on cerebral aneurysm: computational fluid dynamic study of 20 Middle Cerebral Artery Aneurysms. *Stroke*, *35*(11), 2500–2505.

Song, S.-W., Kim, T. H., Lim, S.-H., Lee, K.-H., Yoo, K.-J. & Cho, B.-K. (2014), Prognostic factors for aorta remodeling after thoracic endovascular aortic repair of complicated chronic DeBakey IIIb aneurysms. *The Journal of Thoracic and Cardiovascular Surgery, 148*(3), 925–933.e1.

Spranger, K., Capelli, C., Bosi, G. M., Schievano, S., & Ventikos, Y. (2015). Comparison and calibration of a real-time virtual stenting algorithm using Finite Element Analysis and Genetic Algorithms. *Computer Methods in Applied Mechanics and Engineering, 293*, 462–480.

Spranger, K., & Ventikos, Y. (2015). Which Spring is the Best? Comparison of Methods for Virtual Stenting. *IEEE Transactions on Bio-medical Engineering, 61*(7), 1998–2010.

Stalder, A. F., Frydrychowicz, A., Russe, M. F., Korvink, J. G., Hennig, J., Li, K., et al. (2011). Assessment of flow instabilities in the healthy aorta using flow-sensitive MRI. *Journal of Magnetic Resonance Imaging, 33*(4), 839–846.

Sueyoshi, E. (2004). Growth rate of aortic diameter in patients with type B aortic dissection during the chronic phase. *Circulation, 110*, 256–261.

Svensson, L. G., Kouchoukos, N. T., Miller, D. C., Bavaria, J. E., Coselli, J. S., Curi, M. A., et al. (2008). Expert consensus document on the treatment of descending thoracic aortic disease using endovascular Stent-Grafts. *The Annals of Thoracic Surgery, 85*(1), S1–S41.

Tsai, T. T., Schlicht, M. S., Khanafer, K., Bull, J. L., Valassis, D. T., Williams, D. M., et al. (2008). Tear size and location impacts false lumen pressure in an ex vivo model of chronic type B aortic dissection. *Journal of Vascular Surgery, 47*(4), 844–851.

Tse, K. M., Chiu, P., Lee, H. P., & Ho, P. (2011). Investigation of hemodynamics in the development of dissecting aneurysm within patient-specific dissecting aneurismal aortas using computational fluid dynamics (CFD) simulations. *Journal of Biomechanics, 44*(5), 827–836.

Umaña, J. P., Lai, D. T., Mitchell, R. S., Moore, K. A., Rodriguez, F., Robbins, R. C., et al. (2002). Is medical therapy still the optimal treatment strategy for patients with acute type B aortic dissections? *The Journal of Thoracic and Cardiovascular Surgery, 124*(5), 896–910.

van Bogerijen, G. H. W., Williams, D. M., & Patel, H. J. (2014). TEVAR for complicated acute type B dissection with malperfusion. *Annals of Cardiothoracic Surgery, 3*(4), 423–427.

Verhoeven, E. L., Paraskevas, K. I., Oikonomou, K., Yazar, O., Ritter, W., Pfister, K., et al. (2012). Fenestrated and branched stent-grafts to treat post-dissection chronic aortic aneurysms after initial treatment in the acute setting. *Journal of Endovascular Therapy, 19*(3), 343–349.

Weigang, E., Nienaber, C. A., Rehders, T. C., Ince, H., Vahl, C.-F., & Beyersdorf, F. (2008b). Management of patients with aortic dissection. *Deutsches Arzteblatt International, 105*(38), 639.

Xiang, J., Natarajan, S. K., Tremmel, M., Ma, D., Mocco, J., Hopkins, L. N., et al. (2010). Hemodynamic-morphologic discriminants for intracranial aneurysm rupture. *Stroke, 42*(1), 144–152.

Chapter 5
Role of Vessel Wall Motion in Aortic Dissection

In this chapter, two-way fluid-structure interaction (FSI) simulations are performed and results are compared to rigid wall simulations for the pre-operative case analysed in the previous chapter, in order to evaluate the importance of considering vessel wall and intimal flap motion. To the best of the author's knowledge, this is the first FSI study of AD, which produced wall displacements comparable to clinical imaging data. The study shows that a number of haemodynamic parameters of interest are significantly altered in key regions by the inclusion of wall motion.

5.1 Introduction

The difficulties in selecting the most appropriate treatment for individual patients have led to an increased interest in patient-tailored computational approaches to aid the clinical decision making process; hence a great deal of research has been conducted with the common aim of constructing a framework which could be embedded within vascular clinics to provide clinicians with additional information that is not available from imaging data alone. Recent developments in medical imaging technology, such as 4D MRI, have enabled dynamic measurement of velocities and wall movement. However, further improvements in the technology are required in order to have sufficient resolution to accurately capture WSS and calculate pressure, particularly in the case of AD, where extremely low flow rates can be observed in the FL (Francois et al. 2013).

The work presented in this chapter was published in 'Aortic dissection simulation models for clinical support: fluid-structure interaction versus rigid wall models', *Biomedical Engineering OnLine* (Alimohammadi et al. 2015b).

Patient-tailored CFD approaches provide a promising alternative. However, as with most numerical simulations, these approaches require a number of assumptions to be made, in this case regarding the boundary conditions (pressure or flow, dynamic or static, fixed or moving wall) and flow properties (viscosity model for blood, turbulence).

Studies in healthy aortae have shown that inclusion of the motion of the vessel wall alters the WSS values (Lantz et al. 2014; Reymond et al. 2013). However, it is not clear how much WSS will be affected by the wall motion in simulations of AD, wherein vessel wall distensibility is reduced (Ganten et al. 2009). Several FEM studies have been carried out to investigate the effect of pressure on the stress distribution in the vessel wall in relation to AD (Thubrikar et al. 1999; Nathan et al. 2011). Wall stresses have been indicated as a cause for tear formation or enlargement (Nathan et al. 2011; Thubrikar et al. 1999). Khanafer and Berguer (2009) carried out an FSI study and analysed their results in the context of AD, although the computational geometry was a straight tube, and observed an increased stress within the medial layer (Khanafer and Berguer 2009), which may explain the separation of the medial and intimal layers in AD initiation. It should be noted that none of these studies were carried out on full dissected aortic geometries, but rather on representative geometries or healthy aortae.

Furthermore, only a single study on AD considering wall motion and fluid dynamics has been reported hitherto (Qiao et al. 2014). However, they used a linear elastic model with an extremely high modulus of elasticity of >100 MPa. Although previous studies of the aortic wall have used a range of values from 0.4–6 MPa (Crosetto et al. 2011; Reymond et al. 2013; Colciago et al. 2014; Kim et al. 2009; Xiao et al. 2013; Gao et al. 2006; Brown et al. 2012; Nathan et al. 2011; Khanafer and Berguer 2009) and wall elasticity is decreased in AD, it is unlikely to be decreased by two orders of magnitude, as Ganten et al. (2009) reported an average decrease of only 12%. As a result, despite some rigid body motion of the descending aorta, the motion of the flap relative to the vessel walls was negligible in the study of Qiao et al. (2014). Furthermore, the geometry was excessively smoothed and the branches removed.

Clinical imaging studies have reported the motion of the IF and corresponding contraction and expansion of the two lumina using ECG-gated CT (Ganten et al. 2009; Yang et al. 2014) or 2D pcMRI (Karmonik et al. 2012a). The amount of motion of the IF varies considerably between subjects. Certain individuals in the study of Ganten et al. (2009) had displacements of as much as 7.9 mm across the cardiac cycle, although the median was 1.3 mm from 23 men and 9 women in the age range 46–84. Yang et al. (2014), in a cohort of 33 men and 16 women of ages 30–73, observed greater flap motions with an average of 5.5 ± 2.5 mm (mean $\pm$ standard deviation) and a maximum of 10.2 mm. Karmonik et al. (2012a) analysed 42 patients using MRI, and observed smaller flap motions, approximately in the range 0–1 mm. In contrast to the large displacements observed in some patients, Ganten et al. (2009) observed no flap motion at all in 43% of the patients considered in their study (although half of these had 'healed' dissections), and Chen et al. (2013a) also reported negligible movement. IF motion is thus clearly a highly patient-specific parameter which

5.1 Introduction

may have a considerable effect on predictions of haemodynamics using numerical methodologies.

FSI simulations, which couple CFD simulations of the fluid with FEM of the aortic wall, are capable of capturing this motion (Brown et al. 2012; Reymond et al. 2013; Moireau et al. 2011; Crosetto et al. 2011; Kim et al. 2009) but require specific expertise, are subject to further assumptions and are prohibitively more computationally expensive. This warrants further investigation as to whether the additional effort and resources required to incorporate the wall motion are justified in simulations of AD, in the context of clinical translation. Taking these issues into account, this chapter comprises a preliminary investigation into the application of FSI simulations in the computational analysis of a type-B AD.

5.2 Methods

5.2.1 Geometry

The pre-operative case from Chap. 4 (similar to Chap. 3, but with the aneurysm and bovine arch removed) was used to define the fluid domain. As described previously, it is very difficult to capture the vessel wall thickness directly from CT scans (Erbel et al. 2001; Speelman et al. 2008), due to the limited resolution of images, and the low contrast between the vessel wall and lumen (see Fig. 4.2). Hence, a simple model for the wall geometry was generated by assuming a constant external vessel wall thickness throughout the domain. The fluid geometry was imported into Geomagic Studio (3D systems, Rock hill, USA) and was expanded by 2.5 mm normal to every face. Subtracting the original geometry from the extruded geometry yielded a model for the vessel wall, in which the thickness was uniform for the external wall. The use of a constant wall thickness has been reported in other studies, with 2 mm (Speelman et al. 2008; Qiao et al. 2014) or 2.2 mm (Nathan et al. 2011). Aortic wall thickness varies considerably in the population, and values less than 4 mm are considered normal (Erbel and Eggebrecht 2006). Malayeri et al. (2008) reported that patients with hypertension had a larger mean aortic thickness than normotensive patients, with a value of 2.45 mm. Given that the majority of AD patients are hypertensive (Khan and Nair 2002), the value of 2.5 mm was selected as a representative value for this study.

The intimal flap thickness was automatically produced, based on the gap between the two lumina in the fluid model. The flap thickness was 2.45 ± 0.34 mm (median median absolute deviation). Flap thickness is not generally reported in the literature, precluding evaluation of this range as compared to other studies. As the flap comprises the intima and some of the media layer only, the IF could be expected to initially be thinner than the three layered outer wall of the TL. However, after the dissection occurs, neointimal tissue formation and fibrosis act to thicken the flap and increase its rigidity (LePage et al. 2001). Fig. 5.1 shows the solid and fluid geometries.

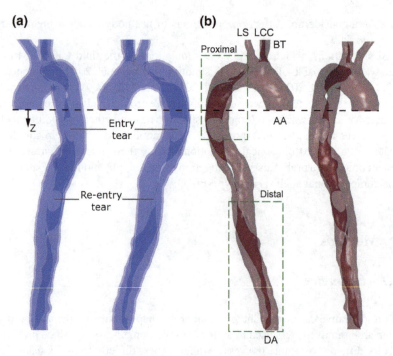

Fig. 5.1 Geometry of **a** Solid **b** fluid. Entry and re-entry tears are indicated, as is the co-ordinate z. The domain boundaries are AA ascending aorta; DA distal abdominal; LS *left* subclavian artery; LCC *left* common carotid; BT brachiocephalic trunk. *Dashed green* boxes show the regions analysed in Figs. 5.9 and 5.10

The fluid and solid domains were meshed together using ANSYS Workbench/ICEM-CFD (ANSYS Inc, Canonsburg, USA), and had approximately 230,000 and 50,000 elements, respectively. For the fluid mesh, 7 prismatic layers were included at the vessel wall. The simulation was run for three cardiac cycles, and a periodic state was reached after two cycles with appropriate initialisation. The third cycle was extracted and used for the proceeding analysis. For the purposes of comparison, an additional simulation was carried out with the same fluid mesh and boundary conditions, but without the solid domain, and with fixed walls. This will be referred to as the 'rigid wall' simulation herein.

5.2.2 Boundary Conditions

The blood was modelled as an incompressible fluid with a density of 1056 kg/m^3. In order to further develop the model presented in Chaps. 3 and 4, the fluid was considered to be non-Newtonian, with viscosity defined by the Carreau-Yasuda model. The Carreau-Yasuda model incorporates zero- and infinite-shear viscosities, μ_0 and

5.2 Methods

Table 5.1 Parameters used for Carreau-Yasuda blood viscosity model (Gijsen et al. 1999) and hyperelastic wall model of (Raghavan and Vorp 2000)

Blood viscosity					Wall hyperelasticity	
μ_0 (m Pa s)	μ_∞ (m Pa s)	a	m	λ_{CY} (s)	A (MPa)	B (MPa)
22	2.2	0.644	0.392	0.110	0.174	1.881

μ_∞, a time constant λ_{CY} and two exponents a and m. The viscosity is stated as a function of shear rate, $\dot{\gamma}$ according to:

$$\mu = (\mu_0 - \mu_\infty)\left(1 + (\lambda_{CY}\dot{\gamma})^a\right)^{(m-1)/a} + \mu_\infty \tag{5.1}$$

The parameter values used in this study were calculated by Gijsen et al. (1999) by fitting the model to viscosity measurements of a blood analogue and are listed in Table 5.1.

Although turbulence was not found to significantly affect the results in Chap. 3, with the additional vessel wall motion and non-Newtonian fluid properties, it seemed worthwhile to include turbulence in the FSI simulations reported in this chapter. The SST model described in Chap. 3 was used. The inlet flow and three-element Windkessel models used in the preceding chapters were applied at the boundaries of the fluid domain. Due to the external coupling between CFX and ANSYS Mechanical required for two-way FSI, the Windkessel models were implemented directly in CFX. The average flow rate was calculated as in Eq. 3.1, and the instantaneous pressure, $P_{0D,i}(t)$ was calculated based on Eq. 2.15, using the pressure and time derivative of the flow rate from the previous time step, $P_{0D,i}(t - \Delta t)$ and $dQ_{3D,i}/dt(t)$ respectively, and the present flow rate, $Q_{3D,i}(t)$. The calculated pressure was passed back to CFX, and was applied as a uniform BC. The time-step used for both FSI and rigid wall simulations was 5 ms.

The vessel wall was modelled using the hyperelastic model of Raghavan and Vorp (2000), using parameters calculated therein, which are given in Table 5.1 (see Eq. 2.22). As shown in Fig. 2.17, this model is similar to the use of a linear elastic material, but displays a small amount of load-stiffening. As well as providing a slightly more accurate representation of the vessel wall mechanics, this model improves computational robustness in ANSYS, which is of particular importance in the context of AD, wherein the IF, with its complex geometry and transluminal pressure gradients, is particularly challenging for convergence.

The geometry extracted from the CT scans represents the aorta under a loaded state (blood pressure 50–100 mmHg), therefore the undeformed geometry will have an associated pre-loaded configuration. Accounting for this is important in accurately predicting the wall stresses (Speelman et al. 2009). A number of methodologies have been used to estimate the pre-stress in the vessel wall (Gee et al. 2010; Speelman et al. 2009; Raghavan et al. 2006; Prasad et al. 2012; Das et al. 2015). However, such methods are time-consuming and do not necessarily yield unique solutions

(Prasad et al. 2012). The implications of a pre-stress can be considered in terms of the 1D tension stress response shown in Fig. 2.17: the effect of a pre-stress would be an effective shift of the stress values to the right; thus, the load stiffening response would be enhanced. Note that for a linear elastic material, pre-stress would have no effect on the deformation. The omission of a pre-loaded state on the present simulation therefore means that the wall response is closer to that of a linear-elastic material than would be observed if pre-load was estimated. However, the process of including a pre-stress would add to computational time/complexity and in the absence of data for confirming the results, would not necessarily improve the accuracy of the simulation.

An external pressure of 52.5 mmHg (the diastolic pressure in the dissected region) was applied. The centre point of each of the solid domain boundaries was fixed, and motion was restricted to the xy-plane only, thereby allowing the boundaries to expand and contract, but restricting rigid body movement of the aorta. In order to achieve numerical stability, small elastic supports were applied to the outside of the aorta. Although the elastic supports slightly restrict the wall motion, the absence of patient-specific wall data means that this is equivalent to an increase in the estimated wall stiffness, which as described above is not well defined, even for healthy aortae.

5.3 Results

5.3.1 Wall Displacement

The displacement of the aortic wall is shown in Fig. 5.2 for three time instances indicated in inset graphs. Figure 5.2a shows the displacement at mid systole. The ascending aorta, particularly around the branches has been displaced outwards, increasing the volume of the aorta. The edges of the entry tear are displaced by up to 0.75 mm and a displacement of the IF distal to the re-entry tear of approximately 0.5 mm can be observed. At peak systole, two views are shown (Fig. 5.2b, c). At this stage in the cardiac cycle, the ascending aorta has expanded further and the area dilation is ≈7%. The IF motion around the tear is further increased to >0.8 mm, and the IF distal to the re-entry tear is deformed considerably. At the dicrotic notch, (Fig. 5.2d), the deformation around the entry tear is similar to that at peak systole, and distal IF has returned closer to its original position.

5.3.2 Cross-Sectional Area

In order to analyse the variation of cross-sectional area of each lumen and how they are affected by the IF movement, an image processing methodology was utilised in MATLAB (Mathworks, Natick, MA, USA) and LabVIEW (National Instruments, Austin,

5.3 Results

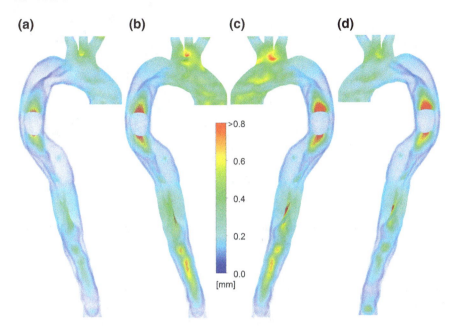

Fig. 5.2 Displacement of the vessel wall at various time instances. Contours show the displacement relative to the undeformed geometry in the *left* posterior view at **a** mid systole, **b** peak systole and in the *right* anterior view, **c** peak systole and **d** dicrotic notch

TX, USA). At each timestep, a video of sequential z-planes along the descending aorta was generated using ANSYS CFD-Post. Each video was imported into Lab-VIEW, wherein an image processing methodology was implemented to identify the areas of the two lumina (TL and FL). The video was generated such that the lumen was black and the background white. Thus each slice was a binary image with one or two (combined lumen or TL and FL) distinct regions, which could be identified using built-in particle identification functions, which also yield the area of each region. Having identified the true lumen, it's centroid was tracked along the stack to ensure that the correct lumen was identified for each slice. The resulting output was the cross sectional area for each lumen, as shown for the undeformed geometry (A_u) in Fig. 5.3. It can be seen that the TL (red) is smaller than the FL for the majority of the descending aorta, until the end of the FL, where the TL becomes the only lumen. At the two tear regions, the area of each lumen is not defined, but the combined cross sectional area increases slightly (at $\approx$20 and $\approx$105 mm) as the space occupied by the IF immediately upstream of the tear is incorporated into the lumen. Between the two tears, the TL is much smaller than the FL.

For each timestep, the cross sectional area of each lumen was calculated along the z-axis. In order to provide a visualisation of the deformation over the cardiac cycle along the descending aorta, the lumina area ratio $A^* = (A_d - A_u)/A_u \times 100\%$, was

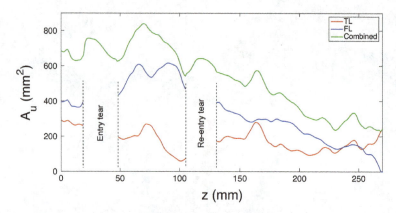

Fig. 5.3 Cross sectional area of each lumen and the combined area along the descending aorta for the rigid wall model, A_u. See Fig. 5.1 for co-ordinate system

defined, where the subscripts d and u represent deformed and undeformed (rigid) geometries respectively.

Figure 5.4a shows A^* as a function of z (defined relative to the inlet: see Fig. 5.1) and t^* (the relative time instance in the cardiac cycle). Positive (red) values of A^* indicate an expansion of the cross sectional area, and negative (blue) values indicate a contraction. Gaps in the A^* maps correspond to the tear regions, wherein TL and FL are not defined. The true lumen cross-sectional area generally decreases as a result of the motion of the IF. Proximal and distal to the entry tear, the deformation of the tear observed in Fig. 5.2 can be seen throughout the cycle, with the greatest area contraction around systole. A region of low deformation exists between the two tears ($z \approx 50$–105 mm), but the IF deforms at the boundary of the re-entry tear, reducing the TL area. In the distal FL, there are large regions of considerable deformation, particularly around $z = 200$ mm at peak systole. At the very distal TL, the cross sectional area increases slightly at systole. As expected, the expansion of the FL is coupled with contraction of the TL, as shown in Fig. 5.4b. The most striking change in cross sectional area takes place in the distal FL, which increases significantly in area over systole, and then relaxes rapidly. It should be noted that due to the tortuosity of the geometry, and the absence of a single luminal centerline from which to establish a normal direction, the A^* maps in Fig. 5.4 are limited to the regions where the axes of the TL and FL are approximately parallel to the z-axis, and thus do not cover the deformation of the proximal FL.

5.3.3 Wall Stress

In order to evaluate the stress on the vessel wall, the von Mises stress is used. This parameter is an estimate of an equivalent stress (relative to the yield stress of the

5.3 Results

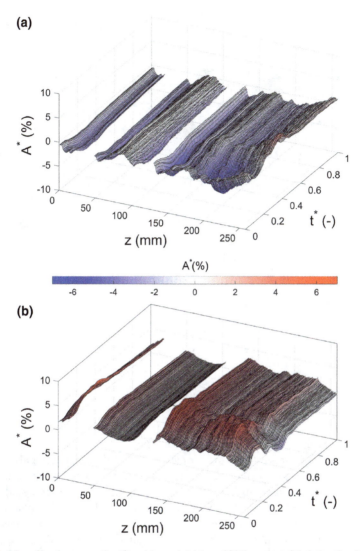

Fig. 5.4 Map of lumina area ratio $A^* = (A_d - A_u)/A_u \times 100\%$, against t^* (relative time in cardiac cycle) and z. **a** True lumen, **b** false lumen

material) and is calculated based on the principal stresses at a given location according to (Khanafer and Berguer 2009)

$$\sigma_{VM} = \sqrt{\frac{(\sigma_1 - \sigma_2)^2 + (\sigma_1 - \sigma_3)^2 + (\sigma_2 - \sigma_3)^2}{2}} \quad (5.2)$$

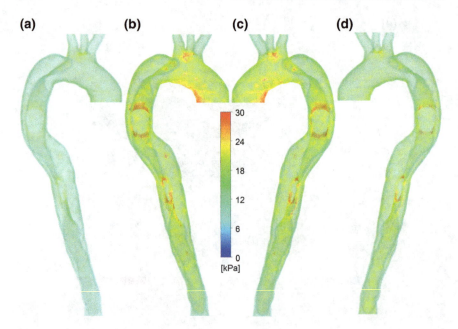

Fig. 5.5 Von Mises stresses in the vessel wall in the *left* posterior view at **a** mid systole, **b** peak systole and in the *right* anterior view, **c** peak systole and **d** dicrotic notch

The von Mises stresses are useful as they represent an estimate of the 3D stress state with a single value for each element.

During mid-systole (Fig. 5.5a), a uniform stress of approximately 12 kPa can be seen along the dissected aorta. Both right anterior and left posterior views of the values of von Mises stress at peak systole can be seen in Fig. 5.5b, c. Highly elevated values can be seen around the lower part of the aortic arch, by the base of the aortic arch branches, around the both tears and proximal FL. Figure 5.5d shows that the stresses are high around the tears, even in the deceleration phase of the cardiac cycle. However, the absolute values of the von Mises stresses shown here are relatively low as compared to other studies (Nathan et al. 2011; Pasta et al. 2013; Khanafer and Berguer 2009). This may be due to the relatively low pressures as a result of the patient's state of anaesthesia, but detailed analysis of the wall stress distribution will only be appropriate once more clinical information is available on wall parameters. Nonetheless, Fig. 5.5 provides qualitative distributions, indicating where higher values are expected to be located.

5.3 Results

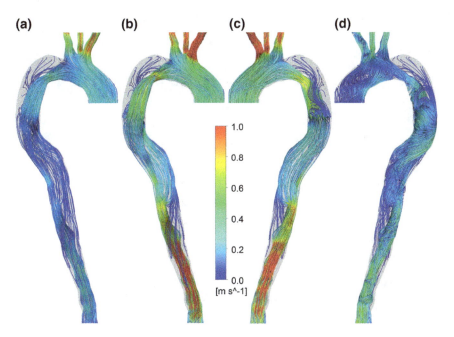

Fig. 5.6 Streamlines in the *left* posterior view at **a** mid systole, **b** peak systole and in the *right* anterior view, **c** peak systole and **d** dicrotic notch

5.3.4 Velocity Distribution

Figure 5.6 shows the streamline patterns at different times in the cardiac cycle. At mid systole (Fig. 5.6a), uniform streamlines can be seen along the aortic arch, branches and along the coarctation region. Around the first tear, the streamlines become less regular, until distal to the re-entry tear in the TL. At peak systole, (Fig. 5.6b, c), uniform streamlines can be observed along the aortic arch and the TL. In the proximal and distal FL, irregular streamlines of low velocity magnitude can be seen. At the dicrotic notch, the streamlines are disordered along the dissected aorta and multiple vortices are present in the aortic arch and along the FL (Fig. 5.6d).

5.3.5 Flow Distribution

The above observations on the wall motion, suggest that the absence of IF motion in the rigid wall simulations is likely to influence the estimated haemodynamic parameters. However, in order to address the question of whether the additional computational effort required for FSI simulations is justified, the significance of the differences in estimated parameters from the two types of simulations needs to be

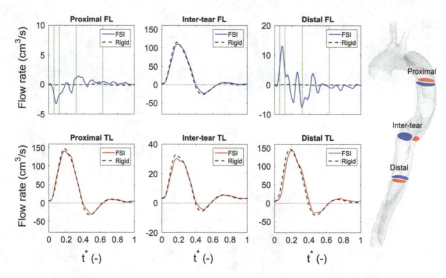

Fig. 5.7 Flow distributions in the descending aorta. Each subfigure compares the flow rate across the indicated plane for FSI and rigid wall simulations against t^*. *Red*—true lumen, *blue*—false lumen. Vertical *green* lines in Proximal FL and Distal FL subfigures indicate time instances considered in Figs. 5.9 and 5.10

established in relation to the clinically useful information required. Figure 5.7 investigates the effects of the IF motion on the proportion of flow going through each lumina at three planes: proximal to the first tear ('Proximal'), between the two tears ('Inter-tear') and distal to the re-entry tear ('Distal'). For each plane, the flow rate through the two lumina obtained from the FSI simulation is shown by solid lines, with the rigid wall simulation results shown by the dotted black lines. In the proximal TL, it can be seen that the shape of the flow waveforms is similar for the FSI and rigid wall simulations, although the compliance of the vessel wall in the former induces a small phase lag in the flow. However, these differences are not expected to be significant in a clinical context. Similar observations can be made regarding the flow in the inter-tear and distal TL. In between the two tears (inter-tear), the majority of the flow is actually going through the FL. This is unsurprising given the relatively larger cross sectional area in this region (Fig. 5.3), but is larger than the 40% of FL flow reported by Chen et al. (2013a).

The flow rates obtained from FSI simulations in the proximal and distal FL are considerably different to those obtained with the rigid wall model; a small (note the different y-axes), but not insignificant amount of flow crosses the selected planes in the false lumen, whereas the rigid wall model predicts almost completely stagnant flow. In the proximal FL, the maximum negative flow occurs at mid-systole, then oscillates between positive and negative flow until about halfway through the cardiac cycle, after which little net flow exists. Note that due to the co-ordinate system defined in Fig. 5.1a, negative flow corresponds to fluid entering the proximal FL. The flow in the distal FL is of higher magnitude than the proximal FL, and is in phase with the proximal FL, i.e. as the flow enters the proximal FL, it also enters the distal FL. The observation that both of these regions expand and contract in phase is similar to that of Rudenick et al. (2013), who observed that fluid entered the FL in systole, and exited in diastole, through both tears in their in vitro phantom.

5.3.6 Pressure Distribution

The pressure distribution is shown at various points throughout the cardiac cycle in Fig. 5.8. Figure 5.8a shows the pressure distribution at mid systole. The pressure is highest at the ascending aorta and decreases at the supraaortic branches and the coarctation. The pressure in the distal FL is higher than the distal TL, whereas in the proximal FL, it is marginally lower than in the adjacent TL. Figure 5.8b, c show the pressure distribution at peak systole. The distribution is similar to mid systole, but the intraluminal pressure gradients are greater. At the dicrotic notch (Fig. 5.8d), the pressure is observed to be lower along the aortic arch and higher in distal TL. Unlike the other two time points shown in Fig. 5.8, the pressure in the proximal FL is slightly higher than the TL (in the same horizontal location) and the pressure in distal FL is lower than the pressure in the distal TL.

The pressure distribution and the streamlines as described do not appear to differ greatly from those observed in the pre-operative case in Chap. 4. Figure 5.7 showed that the two regions of most significant differences were the proximal and distal FL, wherein the movement of the IF significantly altered the flow. The following section considers these regions in more detail.

5.3.7 Proximal and Distal False Lumen

Given that the total net flow over the cardiac cycle in the proximal and distal FL must be zero, there is negligible flow in the proximal and distal FL in the rigid wall simulation (see Fig. 5.7). However, when the wall motion is considered, the area of these regions will vary throughout the cardiac cycle, which results in fluid being drawn in and expelled accordingly, depending on the pressures, inertial forces and wall motion. Figure 5.9 investigates these dynamics for the proximal false lumen, over the region indicated in Fig. 5.1b. In addition to the three time instances shown in Figs. 5.5 and 5.6, a fourth time instance at end diastole is shown, as indicated by the vertical green lines in Fig. 5.7. The right column shows pressure relative to that at the AA (the fluid inlet), so as to illustrate the pressure gradient. The left hand column shows streamlines, which are calculated in both directions (forward and backward) from seed points located on the proximal FL plane (blue) shown in Fig. 5.7, thereby showing the possible source/destination of the flow in the region of interest. At mid systole, a small amount of fluid passing through the coarctation (narrowed region) moves into the proximal FL, although little reaches the most proximal part. At this

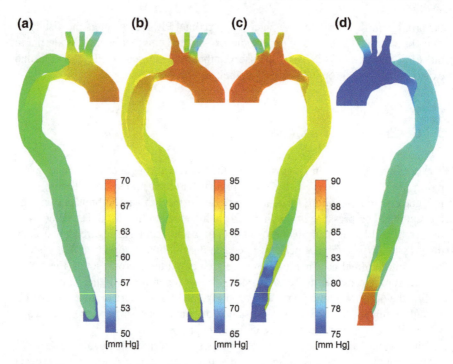

Fig. 5.8 Pressure contours in the *left* posterior view at **a** mid systole, **b** peak systole and in the *right* anterior view, **c** peak systole and **d** dicrotic notch

point, there is a slight pressure gradient between the TL and FL, which may be the driving force for this flow. At peak systole, the pressure gradient is similar and a vortex has developed close to the tear. Interestingly, the separation surface dividing the fluid passing into the proximal FL from the mid TL and FL, has offset to the side of the tear. As in mid systole, there is a small amount of flow into the most proximal part of the FL. At the dicrotic notch, the pressure gradient is inverted, but is of small magnitude (about 5 mmHg). The flow around the tear becomes disordered and the vortex increases in size, extending into the tear region. The small amount of fluid that entered the proximal FL during systole is drawn into the vortex and thus out of the proximal FL. At end diastole, there is negligible fluid movement in the proximal FL, and the flow in the tear region becomes very chaotic. The vortical flow structure around the tear in the proximal FL may explain the oscillations in net flow observed in Fig. 5.7.

Figure 5.10 shows the corresponding distributions for the distal FL (see Fig. 5.1b). The dynamics in this region are significantly different to those in the proximal FL. At mid systole, flow passing through the Inter-tear FL continues directly into the distal FL, and the streamlines extend to the very bottom of the region. At peak systole, there is a large pressure gradient of almost 20 mmHg between the FL and TL, which could be expected to generate considerable flow between the two. Conversely, the

5.3 Results

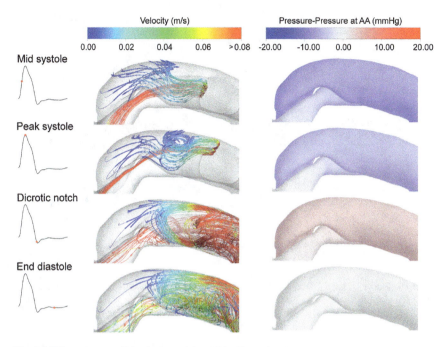

Fig. 5.9 Flow characteristics in the proximal FL. The *left* column shows streamlines calculated from the Proximal plane in the FL (see Fig. 5.7). Each row shows a time instance as indicated by the insets and the vertical *green* lines in Fig. 5.7. The *right* column shows pressure contours relative to the pressure at the AA (Pressure minus Pressure at AA)

flow still enters the distal FL at this time instant, albeit at a relatively low flow rate. At the dicrotic notch, the pressure gradient is inverted, but the fluid in the distal FL is ejected as the pressure decreases and the wall contracts. At end diastole, the velocity in the distal FL is almost zero and the pressures between the lumina are approximately equal.

5.3.8 Wall Shear Stress

It is apparent that these complex dynamics cannot be captured with rigid wall simulations, but the question remains as to whether the additional simulation effort expended in gaining this resolution is necessary. One of the key outputs from these simulations is prediction of the WSS, which is typically analysed using various indices, such as the TAWSS and OSI. The TAWSS shows the average magnitude of the WSS over the cardiac cycle, and its distribution obtained with FSI simulations is shown in Fig. 5.11a. High values of TAWSS can be observed in the supraaortic branches and along the distal TL. In the three branches, there is more flow than would normally

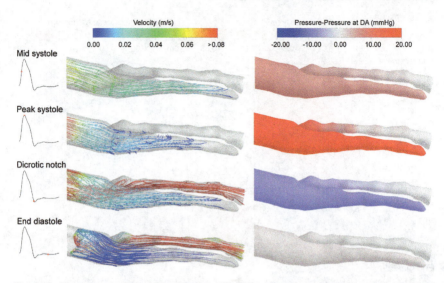

Fig. 5.10 Flow characteristics in the distal FL. The *left* column shows streamlines calculated from the Distal plane in the FL (see Fig. 5.7). Each row shows a time instance as indicated by the insets and the vertical *green* lines in Fig. 5.7. The *right* column shows pressure contours relative to the pressure at the DA (Pressure minus Pressure at DA)

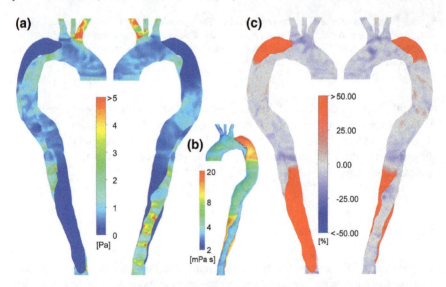

Fig. 5.11 Wall shear stress characteristics. **a** TAWSS distributions for the FSI simulation, **b** dynamic viscosity distribution at peak systole, **c** percentage difference in TAWSS relative to the rigid wall model

5.3 Results

be expected in a healthy aorta, due to the increased hydrodynamic resistance of the pathological thoracic aorta. This results in elevated values of TAWSS in the branches. Despite the reduced flow rate through the distal TL discussed in Chap. 3, the local velocities are increased due to its geometric constriction, and hence elevated TAWSS values are observed. In the proximal and distal FL regions, the TAWSS is very low. Figure 5.11b shows the dynamic viscosity a peak systole. The high values of viscosity in the proximal and distal FL, as predicted by the Carreau-Yasuda model for the shear-thinning properties of blood, may be responsible for reducing velocities in this region, which would decrease WSS. However, the higher viscosity would also act to increase WSS (as WSS is defined as velocity gradient multiplied by local viscosity).

The percentage change in the calculated TAWSS as a result of considering the wall motion in the simulation is shown in Fig. 5.11c. Throughout the TL, a difference of up to -20% can be observed, indicating that the rigid wall model slightly overestimates the TAWSS. In the distal and proximal FL however, a large difference of significantly more than 50% can be seen (colour axes are limited to $\pm 50\%$ for clarity). This is due to the near zero velocity values observed in the rigid wall simulation, which appears to have a notable impact on the TAWSS values.

In absolute terms the differences between the rigid wall and FSI models are not that significant (approximately ± 0.2 Pa), and thus if only TAWSS was of interest, FSI modelling might not be required. However, approximate distributions of TAWSS alone are unlikely to provide sufficient indication of future developments in the patient.

In contrast to the TAWSS, the OSI varies considerably throughout the domain (Fig. 5.12), except in the distal and proximal FL, where it exhibits consistently high values, meaning that the directional changes are large relative to the mean flow. When considered along with Fig. 5.11, it can be observed that these regions appear to exhibit both high OSI and low wall shear stress, which has been identified as a potential index of high risk regions for aneurysm rupture (Xiang et al. 2010; Meng et al. 2014). Figure 5.12b shows the percentage change in the calculated OSI due to wall motion. It can be seen that there are considerable differences in the proximal and distal FL (approximately ± 0.15 in absolute terms), where the complex flow dynamics analysed in Figs. 5.9 and 5.10 significantly enhance the oscillatory nature of the flow.

5.4 Mesh Sensitivity and Efficiency

The key question investigated in this chapter is whether the substantial computational time required to run an FSI simulation of AD is worthwhile in the context of clinical translation. The rigid wall simulation took just 12 h to complete 3 cycles and reach the periodic state, in contrast to the 170 h required for the FSI simulations on a desktop computer (Intel i7, 4 cores, 32 GB RAM). On a workstation (Intel Xeon E5, 8 cores, 32 GB RAM), the simulation time for the FSI model was reduced by approximately 35%. More powerful computing resources could potentially be acquired for clinical

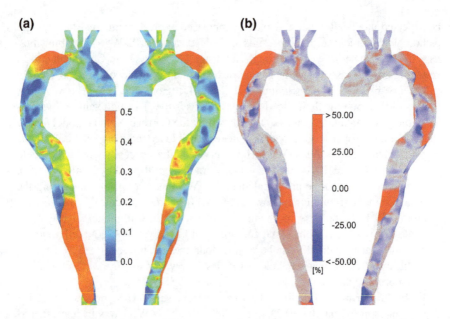

Fig. 5.12 OSI characteristics. **a** OSI distributions for the FSI simulation, **b** percentage difference in OSI relative to the rigid wall model

use, but it is clear that computational efficiency will be an important aspect of the model.

The computational expense of the simulation is directly related to the number of mesh elements. In Chaps. 3 and 4, mesh sensitivity analysis on the fluid mesh demonstrated that a fluid mesh of ≈250,000 elements was sufficient to obtain results that didn't change significantly with further refinement. Although further refinement did slightly alter the results, the difference was small. For coarser meshes, the differences were larger, and so the 'medium mesh' was selected as an appropriate trade-off between accuracy and efficiency. For the FSI simulations, which are inherently more computationally expensive, this trade-off is more critical.

The solid mesh used in the results presented in this chapter had ≈50,000 mesh elements, as FEM generally requires fewer elements than CFD. This solid mesh was combined with a medium fluid mesh as described in Sect. 5.2.1. In order to analyse both the extra computational expense and the sensitivity of the results to further mesh refinement, the geometry was re-meshed to create a mesh with ≈150,000 solid elements and ≈530,000 fluid elements; herein termed the 'fine mesh'. It was not possible to generate a coarser mesh for the solid using ANSYS, due to the complex morphology of the solid domain, and thus only the two mesh densities were considered for the mesh sensitivity analysis. However, it was established in Chaps. 3 and 4 that the medium fluid mesh is appropriate. Sample images of the two solid meshes can be found in Appendix A.

5.4 Mesh Sensitivity and Efficiency

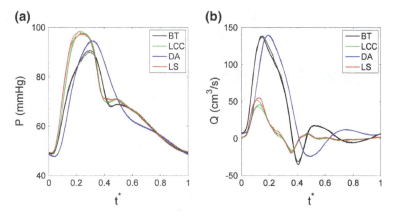

Fig. 5.13 Influence of the mesh refinement on the boundary conditions. *Solid lines* medium mesh, *dashed lines* fine mesh. **a** Pressure at the boundaries, **b** Flow rates at the boundaries

As the boundaries are able to expand and contract in the FSI model, it is important to consider whether the results of the coupling to the Windkessel models at the domain boundaries are affected by the mesh refinement. Figure 5.13a shows the pressure waves at each of the coupled boundaries for the medium (solid lines) and fine (dashed lines) meshes. Small differences can be observed at peak systole in the LCC and LS of up to ≈1 mmHg, with the fine mesh predicting slightly higher pressures. At the maximum pressure in the BT, the fine mesh predicts a slightly lower value, but only by 0.6 mmHg. Elsewhere in the cardiac cycle the deviations between the two meshes are even smaller.

The flow rates at the boundaries (Fig. 5.13b) are also very similar for the two meshes. The largest differences are in the LS and LCC at peak flow, wherein the medium mesh predicts flow rates that are 10% larger. In the BT, the negative flow at the dicrotic notch was slightly greater in the medium mesh, again by ≈10%. Except at these peaks, the differences between the are very small throughout the cycle.

Figure 5.14 shows the wall displacement at peak systole. Figure 5.14a is a duplicate of Fig. 5.2c showing the medium mesh, and is provided as a comparison to Fig. 5.14b, which shows the displacement for the fine mesh. It can be seen that the distribution of the displacements is very similar for the two meshes, although the fine mesh predicts slightly less movement. This can be observed at the tear, the ascending arch and the IF between the distal FL and TL. Figure 5.14c shows the difference in mesh displacement at peak systole. The differences are less than 0.1 mm throughout, with the exception of the arch where the difference in displacement is still less than 0.2 mm.

From this analysis, it could be said that the medium mesh slightly over-predicts the displacement, although the differences are small and may be within the range of error between the FSI predictions of wall motion and the true wall motion experienced by the patient, which would also vary from cycle to cycle.

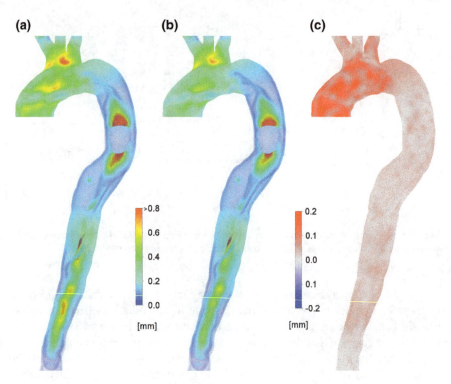

Fig. 5.14 Displacement at peak systole for **a** the medium mesh and **b** the fine mesh. **c** Difference in displacement between the two meshes

The wall shear indices TAWSS and OSI are of potential clinical interest, and thus it must be investigated whether the mesh refinement alters the conclusions that would be drawn based on the WSS. Figure 5.15a, b show the TAWSS distribution for the two meshes. Very minor differences can be seen in the regions of elevated WSS, such as in the LCC and distal TL. For OSI, the region of very high values in the proximal FL is slightly larger for the medium mesh (Fig. 5.15c) than the fine mesh (Fig. 5.15d). On the contrary, the region of high OSI in the distal FL is slightly larger in the fine mesh. Small deviations in the patterns of OSI are also present throughout the domain. However, for practical use in the clinic, the results for TAWSS and OSI with the two meshes are effectively identical.

Figures 5.13, 5.14 and 5.15 showed small differences between estimated parameters using the medium and fine meshes, but these differences are unlikely to lead to different conclusions from the simulation. The simulation time required to run the fine mesh was 2.7 times that required for the medium mesh. It is clear that the significant additional computational expense required does not provide a comparable improvement in the outcome. Therefore, the medium mesh used in the analysis in this chapter is more appropriate for the intended use of the simulation.

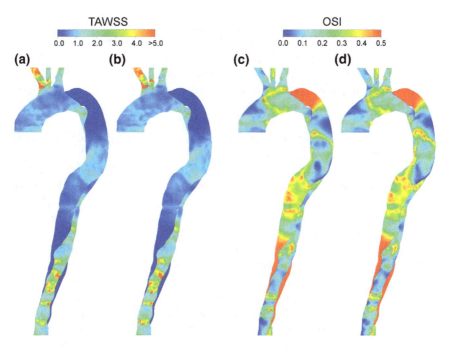

Fig. 5.15 Wall shear stress indices for the two meshes. TAWSS for **a** medium and **b** fine meshes. OSI for **c** medium and **d** fine meshes

5.5 Discussion

The present study investigated whether the additional complexity, and critically additional computational time, required to perform patient-tailored FSI simulations when modelling AD as a tool for interventional planning is justified. The FSI simulation took 14 times longer to run than the rigid wall simulation. Although using a more powerful computer would decrease the simulation time, standard computing tools will still struggle to complete the FSI simulation in less than a few days, which may be too long in the case of AD, in which timely intervention is often critical. There is a growing amount of literature on FSI simulations of aortae (Brown et al. 2012; Reymond et al. 2013; Kim et al. 2009; Crosetto et al. 2011; Moireau et al. 2011, 2013), using various models for the structure of the vessel wall (Xenos et al. 2010; Chandra et al. 2013) in order to capture the movement as effectively as possible. However, there is no consensus on how best to define vessel wall properties from imaging data alone as yet. This difficulty is exacerbated in the case of AD, which is often accompanied by increased vessel wall stiffness (Jong et al. 2014), which may not be evenly distributed throughout the domain. The present simulation uses a simple hyperelastic model extracted from experimental data on an aneurysm, a uniform wall thickness throughout (except the intimal flap), and the additional elastic supports for numerical stability, and as such may not perfectly capture the wall dynamics. However, the results showed deformations quantitatively similar to those measured in vivo by other researchers. The change in cross-sectional area of the TL and FL was calculated from ECG-gated CT scans by Ganten et al. (2009) and they observed a mean TL cross-sectional area reduction (-4.4%), comparable to the A^* values shown in Fig. 5.4a. The median IF oscillation observed by Ganten et al. (2009) was 1.3 mm, while Karmonik et al. (2012a) reported average IF displacements of approximately 0.48–0.68 mm from MRI data. These values compare favourably with the IF deformation shown in Fig. 5.2. In a recent paper, Yang et al. (2014) observed much larger flap motions (1.8–10.2 mm). It is not clear at present why such a discrepancy exists between the values reported by Yang et al. (2014), as compared to previous imaging studies (Ganten et al. 2009; Karmonik et al. 2012a).

As discussed in Chap. 3, the visceral arteries were not included in the present study due to the limited quality of the available imaging data. Inclusion of these vessels in future FSI simulations of AD could provide valuable information on the possible dynamic obstruction that can lead to end-organ malperfusion (Criado 2011).

This study provided high-resolution data, from which a detailed 'map' of cross-sectional area changes was generated. Such a map provides an intuitive and simple way to consider the geometric locations where the vessel wall motion is most significant, and thus has the potential to be a useful clinical tool. 4D MRI data would provide a more rigorous validation of the estimated wall motion. However, such time resolved data were not available for the patient geometry studied here and are not routinely acquired in the clinic; hence the similarities between the simulated values and those reported elsewhere provide confidence that the model employed in this preliminary study is appropriate. The pathophysiology of aortic dissection is not well understood, and thus it is not entirely clear what the most useful data to extract from such FSI simulations is. Longitudinal studies involving a large number of patients are required to identify critical parameters/disease markers and establish benchmarks for clinical use. However, in the absence of such direct information, and given that dissections often form from aneurysms (Erbel et al. 2001), it seems reasonable to assume that the pathophysiology of AD may be similar to aneurysms. Evidence suggests that regions of high WSS are implicated in AD; for example, it has been reported that a reduction in shear stress can minimise the propagation of the dissection (Nordon et al. 2011). Additionally, it has been found that initial tear locations are coincident with regions of maximal pressure or WSS (Wen et al. 2009). Thubrikar et al. (1999) also correlated elevated WSS with sites of intimal tears. The locations of the regions of high TAWSS observed in Fig. 5.11 were not significantly altered in the FSI simulation, compared to the rigid wall. However, a considerable amount of evidence also suggests that in the context of aneurysms, regions of collocated high OSI and low TAWSS are of particular risk of rupture (Xiang et al. 2010), calcification (Malek et al. 1999) or wall thickening (Wen et al. 2009). As reviewed by Meng et al. (2014), in such regions, multifaceted endothelial dysfunction is observed, including increased permeability and 'stickiness' along with inflammatory responses (Malek et al. 1999; Chiu and Chien 2011; Xiang et al. 2010; Meng et al. 2014). The proximal and distal FL in the present simulation exhibit both of these characteristics, indicating that they are the regions of most significant risk.

5.5 Discussion

It was established in this work that the rigid wall model was not capable of accurately capturing the fluid motion in these regions. Furthermore, 4D MRI methods are not yet capable of accurately resolving the slow but complex flows in the FL (Francois et al. 2013). Hence, it can be concluded that numerical tools are currently the best option for analysing haemodynamics in AD, as long as wall motion is included in the simulation. 4D MRI data would, however, enable full validation of the methods used in the present study. Such data could provide wall motion and flows (except in the FL) which could be compared to the numerical results. In order to establish better models for wall properties, ex vivo tissue samples from deceased patients could be mechanically tested and the data incorporated into more complex wall mechanics models (Gasser et al. 2006).

An alternative method for including wall motion in such simulations is to acquire fully time-resolved imaging data and directly prescribe the mesh motion to the simulation, using CFD only. This approach has been used in a number of studies (Lantz et al. 2014, 2015; Torii et al. 2009, 2010; Midulla et al. 2012; Goubergrits et al. 2014), and is reported to improve the match between CFD predictions of velocity and clinical data (Lantz et al. 2014). The main advantage of prescribing the wall motion is that, where suitable imaging data is available, the geometric changes in the aorta across the cardiac cycle (including translation of the aortic root) can be accurately prescribed without requiring assumptions about geometric and material properties of the vessel wall. Additionally, the resulting simulation will have low computational expense (as little as 17% (Lantz et al. 2014)) as compared to a full FSI simulation. This approach is promising for the analysis of AD, and should be investigated when suitable data sets become available. In the long term, if better details on the vessel wall properties (such as thickness, anisotropy, stiffness, particularly for the dissected region) and their relevance to the pathology become available in AD, then FSI models using such information would additionally provide information on stresses in the vessel wall, which can not be provided in simulations where the wall motion is prescribed.

If it transpires that the imaging data of Yang et al. (2014) is more characteristic of true flap motion than previous reports (Ganten et al. 2009; Karmonik et al. 2012a), then it is clear that the loss of accuracy through the rigid wall simplification would be exacerbated. However, further developments are required, particularly in calculating the vessel wall properties of a dissected aorta, before this technology will be able to truly predict the fluid and solid dynamics of this condition. Once such models have been developed and validated, use of anisotropic models combining the vessel wall layers in dissected and non-dissected regions may enable accurate estimates of the internal stresses in the wall of a dissected aorta (Gasser et al. 2006), which could provide further clinical indicators (Nathan et al. 2011). In addition, the complex behaviour of blood viscosity is likely to play a greater role in AD than in other aortic diseases, due to the very low flow regions, as highlighted in Fig. 5.11b. There is no consensus as to what constitutes the best non-Newtonian viscosity model for blood, perhaps due to the absence of high-resolution experimental data of blood in appropriate geometries. However, the Carreau-Yasuda model with parameters according to Gijsen et al. (1999) has been widely cited, and was thus used in this study. The influence of the non-Newtonian viscosity characteristics can be observed in Fig. 5.11b, with reduced viscosity at the vessel walls in the TL (due to high local shear rates) and high viscosity in the proximal and distal FL, due to the very low shear rates. Therefore, it can be concluded that, in addition to FSI, non-Newtonian fluid models should be used in AD simulations.

5.6 Conclusions

The model reported in this chapter comprises, to the best of the author's knowledge, the most advanced model of haemodynamics in a dissected aorta reported in the literature to date, with coupled dynamic BCs, non-Newtonian viscosity, turbulence and wall motion all accounted for. From the comparison with rigid wall simulations, we conclude that in order to develop meaningful information for clinicians dealing with AD, wall motion is a necessary component of the model, without which key regions of interest may not be accurately captured. Despite the additional computational cost, the present study shows that this is an important factor in the model, as well as accurate modelling of blood viscosity and application of appropriate BCs. Further work is required to establish the mechanical properties of dissected aortae in order to take the current model to the next level of accuracy; however, this is also likely to further increase the computational time. It may transpire that a reasonable compromise for the clinic would be to use 4D imaging data to impose the vessel wall motion (Lantz et al. 2014), and analyse the haemodynamics. Comparing such results with full FSI studies in a research setting could establish the viability of this approach.

The final chapter summarises the major findings of this thesis and provides suggestions for future directions for the research.

References

Alimohammadi, M., Sherwood, J. M., Karimpour, M., Agu, O., Balabani, S., & Díaz-Zuccarini, V. (2015b). Aortic dissection simulation models for clinical support: Fluid-structure interaction versus rigid wall models. *Biomedical Engineering Online, 14*, 34.

Brown, A. G., Shi, Y., Marzo, A., Staicu, C., Valverde, I., Beerbaum, P., et al. (2012). Accuracy versus computational time translating aortic simulations to the clinic. *Journal of Biomechanics, 45*(3), 516–523.

Chandra, S., Raut, S. S., Jana, A., Biederman, R. W., Doyle, M., Muluk, S. C., et al. (2013). Fluid-structure interaction modeling of abdominal aortic aneurysms: The impact of patient-specific inflow conditions and fluid/solid coupling. *Journal of Biomechanical Engineering, 135*(8), 081001.

Chen, D., ller Eschner, M. M., von Tengg-Kobligk, H., Barber, D., Bockler, D., Hose, R., et al. (2013a). A patient-specific study of type-B aortic dissection: Evaluation of true-false lumen blood exchange. *Biomedical Engineering Online, 12*, 65.

References

Chiu, J.-J., & Chien, S. (2011). Effects of disturbed flow on vascular endothelium: Pathophysiological basis and clinical perspectives. *Physiological Reviews, 91*(1), 327–387.

Colciago, C. M., Deparis, S., & Quarteroni, A. (2014). Comparisons between reduced order models and full 3D models for fluid-structure interaction problems in haemodynamics. *Journal of Computational and Applied Mathematics, 265*, 120–138.

Criado, F. J. (2011). Aortic dissection: A 250-year perspective. *Texas Heart Institute Journal, 38*(6), 694–700.

Crosetto, P., Reymond, P., Deparis, S., & Kontaxakis, D. (2011). Fluid-structure interaction simulation of aortic blood flow. *Computers and Fluids, 43*, 46–57.

Das, A., Paul, A., Taylor, M. D., & Banerjee, R. K. (2015). Pulsatile arterial wall-blood flow interaction with wall pre-stress computed using an inverse algorithm. *Biomedical Engineering Online, 14*(Suppl. 1), S18.

de Jong, P. A., Hellings, W. E., Takx, R. A. P., Išgum, I., van Herwaarden, J. A., & Mali, W. P. T. M. (2014). Computed tomography of aortic wall calcifications in aortic dissection patients. *PLoS ONE, 9*(7), e102036.

Erbel, R., Alfonso, F., Boileau, C., Dirsch, O., Eber, B., Haverich, A., et al. (2001). Diagnosis and management of aortic dissection task force on aortic dissection, European society of cardiology. *European Heart Journal, 22*(18), 1642–1681.

Erbel, R., & Eggebrecht, H. (2006). Aortic dimensions and the risk of dissection. *British Heart Journal, 92*(1), 137–142.

Francois, C. J., Markl, M., Schiebler, M. L., Niespodzany, E., Landgraf, B. R., Schlensak, C., et al. (2013). Four-dimensional, flow-sensitive magnetic resonance imaging of blood flow patterns in thoracic aortic dissections. *The Journal of Thoracic and Cardiovascular Surgery, 145*(5), 1359–1366.

Ganten, M.-K., Weber, T. F., von Tengg-Kobligk, H., Böckler, D., Stiller, W., Geisbüsch, P., et al. (2009). Motion characterization of aortic wall and intimal flap by ECG-gated CT in patients with chronic B-dissection. *European Journal of Radiology, 72*(1), 146–153.

Gao, F., Guo, Z., Sakamoto, M., & Matsuzawa, T. (2006). Fluid-structure Interaction within a layered aortic arch model. *Journal of Biological Physics, 32*(5), 435–454.

Gasser, T. C., Ogden, R. W., & Holzapfel, G. A. (2006). Hyperelastic modelling of arterial layers with distributed collagen fibre orientations. *Journal of The Royal Society Interface, 3*(6), 15–35.

Gee, M. W., Förster, C., & Wall, W. A. (2010). A computational strategy for prestressing patient-specific biomechanical problems under finite deformation. *International Journal for Numerical Methods in Biomedical Engineering, 26*(1), 52–72.

Gijsen, F., van de Vosse, F., & Janssen, J. (1999). The influence of the non-Newtonian properties of blood on the flow in large arteries: Steady flow in a carotid bifurcation model. *Journal of Biomechanics, 32*, 601–608.

Goubergrits, L., Riesenkampff, E., Yevtushenko, P., Schaller, J., Kertzscher, U., Berger, F., et al. (2014). Is MRI-based CFD able to improve clinical treatment of coarctations of aorta? *Annals of Biomedical Engineering, 43*(1), 168–176.

Karmonik, C., Duran, C., Shah, D. J., Anaya-Ayala, J. E., Davies, M. G., Lumsden, A. B., et al. (2012a). Preliminary findings in quantification of changes in septal motion during follow-up of type B aortic dissections. *Journal of Vacscular Surgery, 55*(5), 1419–1426.e1.

Khanafer Khalil, K., & Berguer, R. (2009). Fluid-structure interaction analysis of turbulent pulsatile flow within a layered aortic wall as related to aortic dissection. *Journal of Biomechanics, 42*, 2642–2648.

Khan, I. A., & Nair, C. K. (2002). Clinical, diagnostic, and management perspectives of aortic dissection. *Chest Journal, 122*(1), 311–328.

Kim, H. J., Vignon-Clementel, I. E., Figueroa, C. A., LaDisa, J. F., Jansen, K. E., Feinstein, J. A., et al. (2009). On coupling a lumped parameter heart model and a three-dimensional finite element aorta model. *Annals of Biomedical Engineering, 37*(11), 2153–2169.

Lantz, J., Dyverfeldt, P., & Ebbers, T. (2014). Improving blood flow simulations by incorporating measured subject-specific wall motion. *Cardiovascular Engineering and Technology, 5*(3), 261–269.

Lantz, J., Renner, J., nne, T. L., & Karlsson, M. (2015). Is aortic wall shear stress affected by aging? an image-based numerical study with two age groups. *Medical Engineering and Physics, 37*(3), 265–271.

LePage, M. A., Quint, L. E., Sonnad, S. S., Deeb, G. M., & Williams, D. M. (2001). Aortic dissection: CT features that distinguish true lumen from false lumen. *American Journal of Roentgenology, 177*(1), 207–211.

Malayeri, A. A., Natori, S., Bahrami, H., Bertoni, A. G., Kronmal, R., Lima, J. A. C., et al. (2008). Relation of aortic wall thickness and distensibility to cardiovascular risk factors (from the multi-ethnic study of atherosclerosis [MESA]). *The American Journal of Cardiology, 102*(4), 491–496.

Malek, A. M., Alper, S. L., & Izumo, S. (1999). Hemodynamic shear stress and its role in atherosclerosis, JAMA. *The Journal of the American Medical Association, 282*(21), 2035–2042.

Meng, H., Tutino, V. M., Xiang, J., & Siddiqui, A. (2014). High WSS or low WSS? complex interactions of hemodynamics with intracranial aneurysm initiation, growth, and rupture: Toward a unifying hypothesis. *AJNR American Journal of Neuroradiology, 35*(7), 1254–1262.

Midulla, M., Moreno, R., Baali, A., Chau, M., Negre-Salvayre, A., Nicoud, F., et al. (2012). Haemodynamic imaging of thoracic stent-grafts by computational fluid dynamics (CFD): Presentation of a patient-specific method combining magnetic resonance imaging and numerical simulations. *European Radiology, 22*(10), 2094–2102.

Moireau, P., Bertoglio, C., Xiao, N., Figueroa, C. A., Taylor, C. A., Chapelle, D., et al. (2013). Sequential identification of boundary support parameters in a fluid-structure vascular model using patient image data. *Biomechanics and Modeling in Mechanobiology, 12*(3), 475–496.

Moireau, P., Xiao, N., Astorino, M., Figueroa, C. A., Chapelle, D., Taylor, C. A., et al. (2011). External tissue support and fluid-structure simulation in blood flows. *Biomechanics and Modeling in Mechanobiology, 11*, 1–18.

Nathan, D. P., Xu, C., Gorman, III, J. H., Fairman, R. M., Bavaria, J. E., Gorman, R. C., et al. (2011). Pathogenesis of acute aortic dissection: A finite element stress analysis. *The Annals of Thoracic Surgery, 91*(2), 458–463.

Nordon, I. M., Hinchliffe, R. J., Loftus, I. M., Morgan, R. A., & Thompson, M. M. (2011). Management of acute aortic syndrome and chronic aortic dissection. *Cardiovascular and Interventional Radiology, 34*(5), 890–902.

Pasta, S., Cho, J.-S., Dur, O., Pekkan, K., & Vorp, D. (2013). Computer modeling for the prediction of thoracic aortic stent graft collapse. *Journal of Vascular Surgery, 57*(5), 1353–1361.

Prasad, A., Xiao, N., Gong, X.-Y., Zarins, C. K., & Figueroa, C. A. (2012). A computational framework for investigating the positional stability of aortic endografts. *Biomechanics and Modeling in Mechanobiology, 12*(5), 869–887.

Qiao, A., Yin, W., & Chu, B. (2014). Numerical simulation of fluid-structure interaction in bypassed DeBakey III aortic dissection. *Computer Methods in Biomechanics and Biomedical Engineering, 18*(11), 1173–1180.

Raghavan, M. L., Ma, B., & Fillinger, M. F. (2006). Non-invasive determination of zero-pressure geometry of arterial aneurysms. *Annals of Biomedical Engineering, 34*(9), 1414–1419.

Raghavan, M. L., & Vorp, D. A. (2000). Toward a biomechanical tool to evaluate rupture potential of abdominal aortic aneurysm: Identification of a finite strain constitutive model and evaluation of its applicability. *Journal of Biomechanics, 33*, 475–482.

Reymond, P., Crosetto, P., Deparis, S., Quarteroni, A., & Stergiopulos, N. (2013). Physiological simulation of blood flow in the aorta: Comparison of hemodynamic indices as predicted by 3-D FSI, 3-D rigid wall and 1-D models. *Medical Engineering and Physics, 35*(6), 784–791.

Rudenick, P. A., Bijnens, B. H., Garcia-Dorado, D., & Evangelista, A. (2013). An in vitro phantom study on the influence of tear size and configuration on the hemodynamics of the lumina in chronic type B aortic dissections. *Journal of Vascular Surgery, 57*(2), 464–474.e5.

Speelman, L., Bosboom, E. M. H., Schurink, G. W. H., Buth, J., Breeuwer, M., Jacobs, M. J., et al. (2009). Initial stress and nonlinear material behavior in patient-specific AAA wall stress analysis. *Journal of Biomechanics, 42*(11), 1713–1719.

Speelman, L., Bosboom, E. M. H., Schurink, G. W. H., Hellenthal, F. A. M. V. I., Buth, J., Breeuwer, M., et al. (2008). Patient-specific AAA wall stress analysis: 99-percentile versus peak stress. *European Journal of Vascular and Endovascular Surgery, 36*(6), 668–676.

Thubrikar, M. J., Agali, P., & Robicsek, F. (1999). Wall stress as a possible mechanism for the development of transverse intimal tears in aortic dissections. *Journal of Medical Engineering & Technology, 23*(4), 127–134.

Torii, R., Keegan, J., Wood, N. B., Dowsey, A. W., Hughes, A. D., Yang, G.-Z., et al. (2009). The effect of dynamic vessel motion on haemodynamic parameters in the right coronary artery: A combined MR and CFD study. *The British Journal of Radiology, 82*(1), S24–S32.

Torii, R., Keegan, J., Wood, N. B., Dowsey, A. W., Hughes, A. D., Yang, G.-Z., et al. (2010). MR image-based geometric and hemodynamic investigation of the right coronary artery with dynamic vessel motion. *Annals of Biomedical Engineering, 38*(8), 2606–2620.

Wen, C.-Y., Yang, A.-S., Tseng, L.-Y., & Chai, J.-W. (2009). Investigation of pulsatile flowfield in healthy thoracic aorta models. *Annals of Biomedical Engineering, 38*(2), 391–402.

Xenos, M., Rambhia, S. H., Alemu, Y., Einav, S., Labropoulos, N., Tassiopoulos, A., et al. (2010). Patient-based abdominal aortic aneurysm rupture risk prediction with fluid structure interaction modeling. *Annals of Biomedical Engineering, 38*(11), 3323–3337.

Xiang, J., Natarajan, S. K., Tremmel, M., Ma, D., Mocco, J., Hopkins, L. N., et al. (2010). Hemodynamic-morphologic discriminants for intracranial aneurysm rupture. *Stroke, 42*(1), 144–152.

Xiao, N., Alastruey, J., & Alberto Figueroa, C. (2013). A systematic comparison between 1-D and 3-D hemodynamics in compliant arterial models. *International Journal for Numerical Methods in Biomedical Engineering, 30*(2), 204–231.

Yang, S., Li, X., Chao, B., Wu, L., Cheng, Z., Duan, Y., et al. (2014). Abdominal aortic intimal flap motion characterization in acute aortic dissection: Assessed with retrospective ECG-gated thoracoabdominal aorta dual-source CT angiography. *PLoS ONE, 9*(2), e87664.

Chapter 6
Conclusions and Future Work

6.1 Introduction

The main aim of the present study was to develop and analyse computational methods in the simulation of aortic dissection.

Information on haemodynamic parameters such as pressure, velocity and WSS can be advantageous in approaching the clinical question of how to treat patients with type-B AD. These haemodynamic parameters are very difficult to acquire in vivo due to the tortuous morphological aspects of the disease and the limitations of current imaging or invasive measurements techniques. In AD, the intertwined luminae and near-stagnant flow in parts of the FL limit the applicability of clinical imaging techniques such as 4D MRI. Resolving WSS from 4D MRI data is possible, but is limited in spatial resolution and accuracy. Additionally, even invasive pressure measurements, which are usually acquired as part of the routine clinical intervention, are often unable to provide data in the FL or associated aneurysms, due to the severe risk of iatrogenic trauma.

In recent years, CFD has shown promise as a tool for clinicians, in order to assist them with the decision making process. CFD may be particularly useful in the case of AD, wherein detailed information on haemodynamics are required in critical locations such as the proximal and distal FL, or around an often associated coarctation.

However, such CFD simulations must achieve both sufficient accuracy (reproduction of the true haemodynamic environment) and computational efficiency to achieve computational results within the limited time permitted in the clinical setting.

In this thesis, clinical imaging data and invasive pressure measurements from a patient suffering from type-B AD were used in conjunction with numerical techniques in order to predict the haemodynamic environment in the dissected aorta. The emphasis was on the application of dynamic, patient-tuned outlet BCs and the influence and practicability of modelling vessel wall motion. Analysis of various treatment options was also considered, showing the utility of such models.

6.2 Main Contributions

The main contributions of the present study are as follows:

- The application of appropriate BCs is essential in modelling the aorta. In AD simulations, the applied BCs have always been defined a priori, which has a significant influence on the simulated haemodynamics. In order to address this issue, Windkessel BCs were used in the present study. There is no single established method for defining Windkessel parameters, and the present study proposed an iterative tuning approach, which can be used with maxima and minima of pressure waves recorded invasively. This enabled the first simulation of AD with dynamic BCs and the simulated pressures compared favourably with invasively measured ones. The results thus provided improved predictions of other haemodynamic variables.
- The ability to analyse effects of different treatment options is one of the significant advantages of numerical modelling in aortic disease. However, by changing the geometry, the flow at the boundaries of the domain is also likely to change. Coupling to dynamic BCs allowed for an improved representation of the effects of treatment options in the context of the aorta being a single part of a dynamic system. The results showed a considerable difference between the treatment options considered.
- The movement of the aortic wall is critical to the functioning of the vasculature, as it creates the Windkessel effect, ensuring a continuous supply of blood to the peripheral vessels. The oscillatory expansion and contraction of the aortic geometry is likely to have a notable influence on haemodyanmics. In AD, wherein the IF separates two lumina of different pressures, the influence of the vessel wall is perhaps even more important than in other aortic pathologies. The first FSI simulation of AD with appropriate wall properties, including dynamic BCs, was conducted in this study. The results indicated interesting dynamics in key regions that were not captured by rigid wall models.

6.3 Summary of Main Findings

The accuracy and thus utility of simulations of the blood flow in the aorta are entirely dependent on the modelling assumptions made. These include morphological aspects, such as inclusion/exclusion of supraaortic or visceral vessels, fluid aspects, such as non-Newtonian viscosity and turbulence, and numerical aspects such as domain discretisation and solver preferences. Application of appropriate BCs is also of fundamental importance. The present study aimed to address a number of these issues with a view to developing tools for use in a clinical setting: as such both efficacy and efficiency were considered.

Data from a single patient was used throughout this study. CT scans of the patient were used to construct a 3D geometry representative of the patient's dissected aorta. The supraaortic branches were included in the geometry. No flow data was available

6.3 Summary of Main Findings

for the patient, and so a flow rate from another patient suffering from type-B AD was used in all studies, scaled to the patient's heart rate. Invasive pressure measurements at various locations along the aorta were recorded in the patient as part of the standard clinical procedure, from which the minima and maxima were noted.

The first part of the study entailed the use of the invasive pressure data for tuning dynamic BCs. Based on the limited data, 0D models of the vasculature were selected. Three-element Windkessel models were chosen for their improved ability to capture high frequency characteristics, as compared to the simpler two-element models. The three-element Windkessel model is analogous to an RCR circuit, in which a parallel resistance and capacitance represent the resistance and compliance of the downstream vasculature, in series with another resistor in series to represent the characteristic impedance of the system. Thus at each of the the domains four outlet boundaries (at the supraaortic branches and the distal abdominal aorta), three parameters had to be derived. However, as the system was fully coupled, changing parameters at one outlet would influence the results at the other outlets. To account for this, an iterative multi-stage approach was developed, whereby the parameters were estimated from an initial constant pressure simulation. A 'brute force' approach was then used to estimate the parameters that would yield the desired pressure for a given flow wave. These were then implemented in the 3D model and the process was repeated until further iterations did not improve the match with the invasive measurements.

The maxima and minima of the simulated pressure waves were compared to the invasive measurements and were found to have errors of less than 5% of the pulse pressure, with the exception of the minimum at the inlet at 7%. Furthermore, the simulated pulse pressures were all within 2.5% of the measured range, which is on the order of the uncertainty in the catheter measurements. The sensitivity of the parameters was analysed by altering the invasive pressure measurements by ± 1 mmHg and retuning the Windkessel parameters. The RMS differences in the flow rate were on the order of 0.1% in all branches, indicating that the solution was robust.

The flow rates through the domain predicted by the simulation yielded interesting results. Only 45% of the flow passed through the descending aorta (as opposed to > 80% in a healthy aorta), indicating a risk of lower limb malperfusion syndrome. Additionally, 70% of the flow passed through the FL, rather than the TL. Interesting flow patterns were observed as a result of the pathological morphology of the aorta, and counter-rotating vortices were observed in the aneurysm. The TAWSS was observed to be low in the distal and proximal FL, as was the OSI, with elevated regions of both indices in scattered regions throughout the arch and TL and FL between the tears. Using a coarser mesh was observed to significantly alter the results, particularly pressure and OSI, but further refinement did not make a significant difference. Analysis of the influence of turbulence was carried out, and it was found that the turbulence was low throughout, as the small regions of elevated turbulence intensity ($\approx 25\%$) corresponded to regions of very low absolute velocity.

Having established a model for improved simulation of the haemodynamics in AD, it was utilised in a 'virtual treatment' simulation, in which two options for surgical intervention were investigated by altering the morphological characteristics of the aorta. If predefined BCs had been used, the influence of the coupling between the aorta and the rest of the vasculature would not have been predicted. The use of Windkessel BCs with derived parameters provided a better estimation of the effects that the treatments had.

The flow rates through the branches were observed to change in response to the virtual treatments of a single-stent blocking the entry tear only, or a more extreme operation blocking the re-entry tear and simulating full recovery of the DA lumen diameter (double-stent). The velocities became increasingly smoothened by the operations, and the pressure gradient across the aorta was reduced. In order to quantify the change in the resistance of the morphology to flow, the kinetic energy loss across the cardiac cycle was calculated. Whilst the single-stent operation only reduced the KE loss by 1.5%, the reduction in the double-stent case was > 40%. An analysis of the histograms of WSS indices throughout the domain provided a novel way to analyse the effects of the treatments on the aortic wall. The two surgeries were observed to decrease the area with extremely low TAWSS and increase the symmetry in the OSI distribution. The index TAWSS × (0.5-OSI), which may be associated with a pathological environment for the endothelium, showed that the virtual double-stent would almost entirely remove such areas of elevated risk. Overall the results suggested that the more invasive operation would be significantly more effective than blocking the entry tear only.

In the final part of the study, the influence of wall motion was included in the simulation. When modelling the motion of the wall, two-way coupled FSI simulations are necessary. Thus at each timestep, both the fluid and solid domains must converge and the interfaces must be appropriately matched. This incurs a significant computational cost, which is of particular importance if such simulations are to be translated to the clinic. The FSI simulations took 14 times as long as rigid wall simulations, however the results indicated that the extra cost is necessary. Turbulence was modelled with the SST model, and non-Newtonian viscosity was simulated using the Carreau-Yasuda model. The vessel wall was modelled as hyperelastic.

Wall displacements of almost 1 mm were observed, with a maximum along the intimal flap adjacent to the tears. The observed displacements compared well with previous imaging studies of IF motion. The cross sectional area of the two lumina in the descending aorta was investigated using an image processing based technique to extract data from videos of the numerical results. A 'map' of the area lumina ratio, the proportional change in cross sectional area for each lumen, was given in time and space throughout the cardiac cycle. Analysis of the flow rates through sections of the descending aorta revealed that in the distal and proximal FL, there is a highly oscillatory flow which was almost entirely missed in the rigid wall model. Close inspection of the pressure and streamlines in these regions showed that the movement of the IF was generating flows in the opposite direction to the pressure gradient. The influence of this oscillatory flow on TAWSS was small, as the magnitudes were low, but there was a striking difference in the OSI. Whereas for the rigid wall model the OSI in the distal and proximal FL was very low, in the FSI model the OSI was almost 0.5. This corresponded to regions of extremely low TAWSS; a potential indicator for further damage to the diseased aorta. Additionally, the viscosity in these regions was observed to be elevated, a characteristic which highlights the importance of a non-Newtonian viscosity model and indicates that further development of blood viscosity models for CFD would be advantageous.

6.4 Significance of this Study

This study has presented a number of important developments to the modelling of AD. Direct comparison between the Windkessel and static pressure BCs showed that the flow distribution in the latter was not well modelled, particularly in the DA throughout diastole wherein the flow remained positive throughout. However, the constant pressure BC approach appears to be superior to the flow split BCs, which results in highly unrealistic pressures required to drive peak systolic flow across all boundaries simultaneously. Tuning of Windkessel parameters to invasive measurements is not possible where such data are unavailable, although using ultrasound flow data may provide an alternative. Nonetheless, it is probable that even simple tuning of two-element Windkessel BCs to representative values would be an improvement over predefinition of the BCs.

The present study used a number of analysis techniques, such as the lumina area ratio, kinetic energy loss, and WSS histograms, which have not previously been used in the analysis of AD simulations. The techniques provide a new way of interpreting the simulation results and highlight the potential of numerical modelling for providing unique analyses.

The present study has shown that the motion of the IF has significant impact on the results of AD simulations, and despite the greatly increased computational expense, future studies should endeavour to include wall movement.

6.5 Future Work

This work represents a number of advances towards the final goal of building numerical and simulation tools for clinical use. In the process of implementing and analysing the present developments, a number of key points have been identified as worthy of further investigation.

With regards to fluid dynamics, the fundamental questions of turbulence and viscosity still need to be resolved. For turbulence, initial studies have been carried out comparing, for example simulations with the SST Tran model to MRI velocity measurements. However, further work is required to carry out such comparisons with dynamic BCs and moving walls, so as to separate the effects of these assumptions from the influence of turbulence. The viscosity characteristics of blood also require further elucidation, as the relative efficacy of various empirical shear-thinning models reported in the literature, such as the Carreau-Yasuda and Quemada model, has not yet been established. The in vitro models of AD used by a number of researchers could perhaps be used in conjunction with actual blood and compared to water and non-Newtonian analogues in order to provide data which could be used for validation of numerical models.

The dynamic BCs used in the present study were based on invasive pressure measurements. However, such measurements may not always be available. It will thus be necessary to ascertain whether such approaches are applicable based on flow measurements using ultrasound, or velocity extracted from pcMRI data. Furthermore, the way that the derived parameters vary after intervention needs to be investigated with longitudinal studies.

The vessel wall motion captured in the present study compared well with clinical imaging studies. However, in order to further develop this modelling approach, patient-specific wall properties will be required. One approach to this is to use 4D MRI data and a reverse FE technique to tune the wall properties based on the measured wall motion. This technique would be time consuming, but may be able to yield data on patients prior to operations and not only improve haemodynamics simulations, but also indicate regions of risk in the vessel wall. A complementary approach might be to carry out ex vivo experiments in cadaverous dissected aortae in order to establish general properties of dissected aortae. Such data would also be useful for numerical model validation, as the BCs could be accurately regulated.

Finally, in order to truly establish the utility of numerical models of AD, large scale, long-term studies, in which multiple patients are monitored over long time periods, will be required. Such data would allow hypothesis testing of various indicators of risk, such as the WSS indices associated with high risk in aneurysms. Furthermore, it would allow for development and demonstration of the robust tools, in order to generate trust within the clinical community.

There is still much work to be done in order to establish widespread numerical modelling as a surgical planning tool, but the increasing interest in the approach promises to deliver real advances, which ultimately will improve the outcome for patients suffering from aortic dissection.

Appendix A
Sample Mesh Images

This appendix provides sample mesh images for the meshes used in Chaps. 3 and 5. Meshes for Chap. 4 were similar to those in Chap. 3, so are omitted for brevity (Figs. A.1, A.2 and A.3).

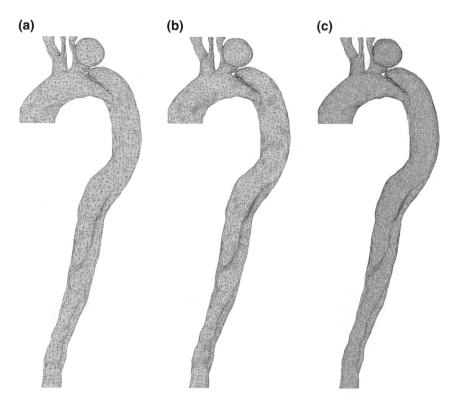

Fig. A.1 Images of the meshes used in the mesh sensitivity analysis for Chap. 3. **a** Coarse, **b** medium, **c** fine

© Springer International Publishing AG 2018
M. Alimohammadi, *Aortic Dissection: Simulation Tools for Disease Management and Understanding*, Springer Theses, https://doi.org/10.1007/978-3-319-56327-5

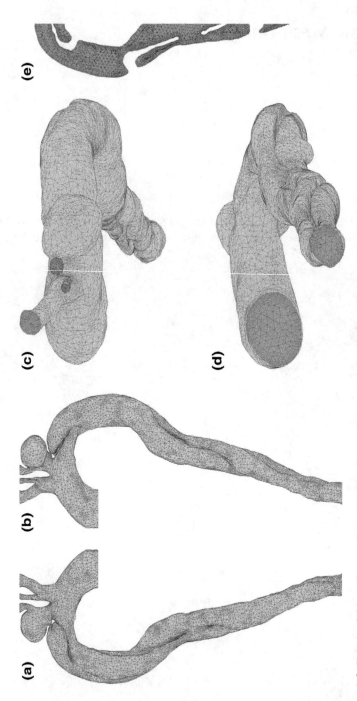

Fig. A.2 Images of the medium mesh used in Chap. 3. **a** Left posterior view, **b** right anterior view. **c** From above, showing the domain boundaries at the supraortic branches, **d** from below, showing the domain boundaries at the ascending aorta and distal abdominal aorta. **e** Cross sectional slice through the mesh

Appendix A: Sample Mesh Images

Fig. A.3 Images of the solid mesh used in Chap. 5. **a** Medium mesh, **b** fine mesh

Appendix B
Detailed Mesh Sensitivity Analysis for Virtual-Stenting Simulations

This appendix provides the full details of the mesh sensitivity analysis for Chap. 4.

Pressure and Flow

Figure B.1a, b shows the flow and pressure respectively at the inlet of all pre-operative cases. As can be seen in Fig. B.1a, there was no significant difference in the flow waves between the three meshes. This is as expected, as the inlet flow wave is prescribed, but confirms that the discretisation at the inlet for the coarse mesh does not introduce error at the inlet. For the pressure waves (Fig. B.1b), the medium and fine mesh have negligible differences, whereas the coarse mesh shows slightly lower pressure values after the peak systolic flow and around the dicrotic notch.

Figure B.1c–f show equivalent comparisons for the single- and double-stent geometries respectively. Similarly to the pre-operative case, the flow at the inlet for both virtually stented geometries is unaffected by the mesh size.

There is a slight difference between the pressure wave for the coarse mesh compared to the other two meshes for the single-stent case (as observed for the pre-operative case). For the double-stent case (Fig. B.1f), there were no significant differences between the three meshes for pressure.

Figure B.2a, c and e, show the flow entering the BT branch for all three different meshes. In all three cases, the medium and fine mesh are very close to each other. Figure B.2b, d and f, show the pressure waves of coarse, medium and fine meshes, for the pre-operative, single-stent and double-stent cases, respectively. In both pre-operative (Fig. B.2b) and single-stent (Fig. B.2d) cases, the pressure between the three mesh cases behaves similarly, with slightly lower pressures around systole and the dicrotic notch as the number of mesh elements decreased. In the double-stent case, the same general trend was observed, but a lower pressure can be seen for the double-stent case throughout the systolic phase.

Figure B.3a, c and e, show the differences in flow between the coarse, medium and fine mesh for the LCC branch. For the double-stent geometry, the coarse and medium meshes have identical flow waves, whereas a slight increase in flow at peak systole was observed for the fine mesh. Similar trends can be observed for the single-stent case. In the pre-operative case, a greater difference was observed between the three

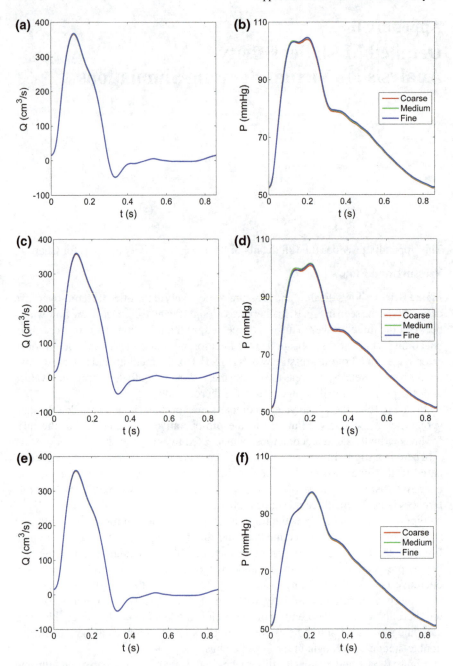

Fig. B.1 **a** Flow and **b** pressure at the AA for all three meshes for the pre-operative case. **c** Flow and **d** pressure at the AA for all three meshes for the single-stent case. **e** Flow and **f** pressure at the AA for all three meshes for the double-stent case

Appendix B: Detailed Mesh Sensitivity …

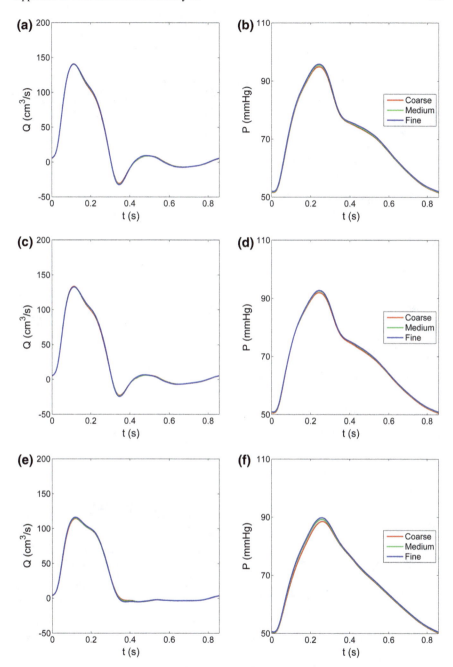

Fig. B.2 **a** Flow and **b** pressure at the BT for all three meshes for the pre-operative case. **c** Flow and **d** Pressure at the BT for all three meshes for the single-stent case. **e** Flow and **f** pressure at the BT for all three meshes for the double-stent case

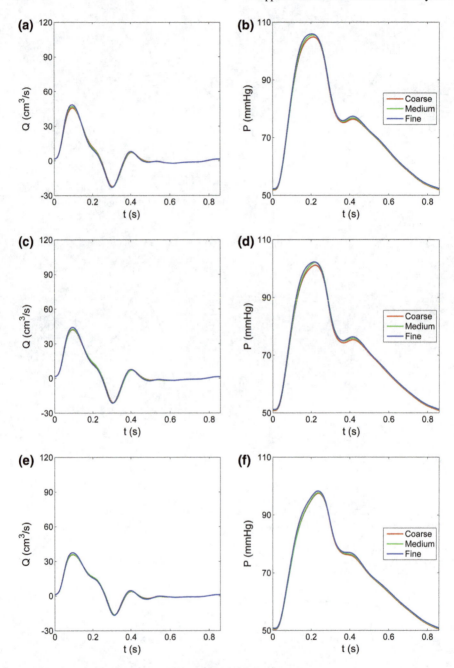

Fig. B.3 **a** Flow and **b** pressure at the LCC for all three meshes for the pre-operative case. **c** Flow and **d** pressure at the LCC for all three meshes for the single-stent case. **e** Flow and **f** pressure at the LCC for all three meshes for the double-stent case

meshes, although the differences are still very small. Differences in pressure for all three cases (Fig. B.3b, d and f, can be observed when the flow is at its maximum and minimum values.

Figure B.4a, c and e, show the flow entering the LS branch for all three cases with their different meshes. There is no significant difference between the coarse, medium and fine mesh in the pre-operative case (Fig. B.4a) and the double-stent case (Fig. B.4e). However, there is a small deviation between the pressure waves for the coarse mesh, compared to the other two mesh types in the single-stent case (Fig. B.4b). All of the calculated pressure waves in the LS branch are in agreement, irrespective of the mesh density (Fig. B.4b, d and f), with the exception of the coarse mesh, for which small differences are observed.

Figure B.5a, c and e show the flow waves and Fig. B.5b, d and f show pressure waves at the DA branch for all three cases and their different meshes. Both the flow and pressure in this branch are effectively the same for all three meshes.

Figures B.6, B.4 and B.8 show contours of the difference between the velocities obtained with the different meshes. For all figures, (a) and (b) show the difference between the medium and fine meshes, and (c) and (d) show the difference between the coarse and medium meshes.

Figure B.6a, b show the difference in velocity between the medium and fine meshes for the pre-operative case at peak systole and at the dicrotic notch, respectively. In Fig. B.6a, the most significant differences ($\approx$0.1 m/s) can be seen along the three ascending aortic branches, around the coarctation and along the distal TL. These differences are reduced at the dicrotic notch (Fig. B.6b), although, around the first tear, the difference is slightly increased at this time instance (the medium mesh velocity is marginally lower than the fine mesh).

Comparing the coarse and medium meshes (Fig. B.6c), differences in velocity can be seen at the domain entry (AA), through the ascending branches, at the coarctation and along the distal TL. Comparing to Fig. B.6a, it can be seen that the differences between the coarse and medium meshes are more pronounced than for the medium and fine meshes. At the dicrotic notch, the magnitude of the velocity differences is decreased. Note that for all three meshes at both points in the cardiac cycle shown here, the velocities in the FL are approximately equal.

The differences in velocity for the single-stent case are shown in Fig. B.7. The major differences between the medium and fine meshes can be seen through the ascending aortic branches and distal TL for the peak systolic case (Fig. B.7a). The velocity differences along the single-stent domain at the dicrotic notch are reduced to very small values (Fig. B.7b). The differences between coarse and medium meshes are also increased through the three branches of BT, LCC and LS at the peak systole (Fig. B.7c). Note that greater differences are observed than between the medium and fine cases at the same time instance (peak systole). This is also the case at the dicrotic notch.

Overall, comparing Figs. B.6 and B.7, it can be seen that the stenting operation reduces the influence of mesh refinement, as might be expected for a less tortuous geometry. This trend is further continued for the double-stent case in Fig. B.8. The velocity differences between the medium and fine meshes through the domain are

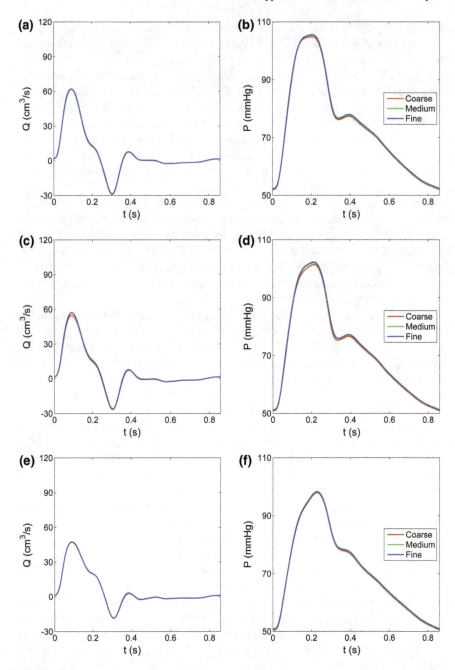

Fig. B.4 **a** Flow and **b** pressure at the LS for all three meshes for the pre-operative case. **c** Flow and **d** pressure at the LS for all three meshes for the single-stent case. **e** Flow and **f** pressure at the LS for all three meshes for the double-stent case

Appendix B: Detailed Mesh Sensitivity … 171

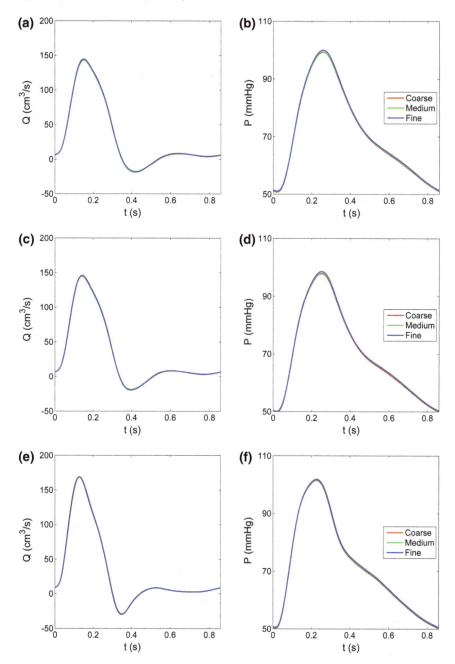

Fig. B.5 **a** Flow and **b** pressure at the DA for all three meshes for the pre-operative case. **c** Flow and **d** pressure at the DA for all three meshes for the single-stent case. **e** Flow and **f** pressure at the DA for all three meshes for the double-stent case

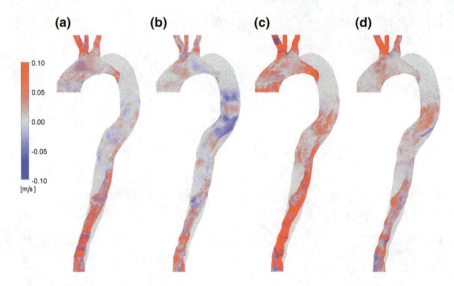

Fig. B.6 Pre-operative case velocity differences: medium and fine meshes at **a** peak systole and **b** dicrotic notch, coarse and medium meshes at **c** peak systole and **d** dicrotic notch

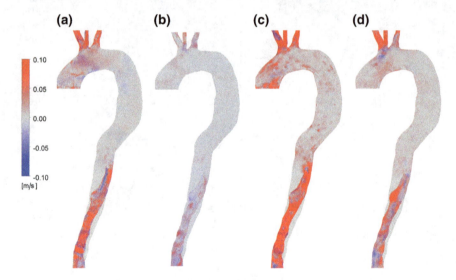

Fig. B.7 Single-stent case velocity differences: medium and fine meshes at **a** peak systole and **b** dicrotic notch, coarse and medium meshes at **c** peak systole and **d** dicrotic notch

Appendix B: Detailed Mesh Sensitivity ...

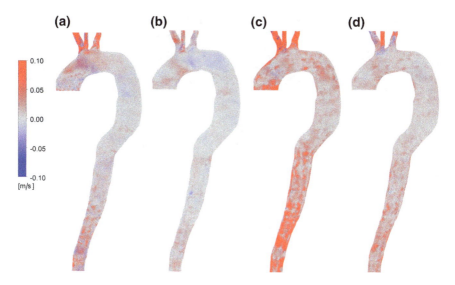

Fig. B.8 Double-stent case velocity differences: medium and fine meshes at **a** peak systole and **b** dicrotic notch, coarse and medium meshes at **c** peak systole and **d** dicrotic notch

further reduced at the peak systole case for the double-stent (Fig. B.8a) compared to the velocity differences observed for the pre-operative and single-stent cases. The only significant differences at peak systole can be seen through the supraaortic branches (which due to their small size are expected to be the most sensitive to mesh refinement and also observe the highest absolute velocities in the domain). At the dicrotic notch, a similar pattern can be observed, but with smaller magnitude (Fig. B.8b). The velocity differences between the coarse and medium mesh for the double-stent case (Fig. B.8c) are higher in magnitude compared to the differences between the medium and fine meshes, but are reduced relative to the single-stent case. This is also the case at the dicrotic notch (Fig. B.8d). It should be noted that these local differences are generally less than 10% of the calculated velocity (compare with Fig. 4.7).

Figures B.9, B.10 and B.11 show contours of the difference between the pressures for the different meshes. For all figures, (a) and (b) show the difference between the medium and fine meshes, and (c) and (d) show the difference between the coarse and medium meshes.

Figure B.9 shows the pressure differences for the pre-operative case at peak systole and the dicrotic notch. At peak systole (Fig. B.9a), the largest differences can be seen along the three branches and the distal TL (≈ -1 mmHg). Also, a difference of ≈ 0.5 mmHg can be seen along the dissected aorta. For the dicrotic notch (Fig. B.9b), the overall differences are reversed but reduced in magnitude (i.e. the fine mesh had marginally higher velocity than the medium mesh).

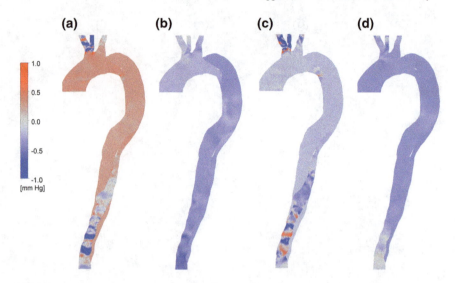

Fig. B.9 Pre-operative case pressure differences: medium and fine meshes at **a** peak systole and **b** dicrotic notch, coarse and medium meshes at **c** peak systole and **d** dicrotic notch

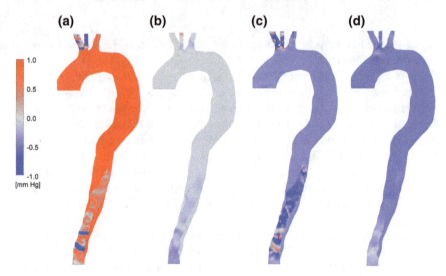

Fig. B.10 Single-stent case pressure differences: medium and fine meshes at **a** peak systole and **b** dicrotic notch, coarse and medium meshes at **c** peak systole and **d** dicrotic notch

Comparing the coarse and medium meshes, (Fig. B.9c, d), the pressure was slightly higher in the medium mesh than in the fine mesh throughout the domain, except in scattered spots in the BT, LCC and distal FL. Again, at the dicrotic notch, there was a small difference between the two meshes.

Appendix B: Detailed Mesh Sensitivity … 175

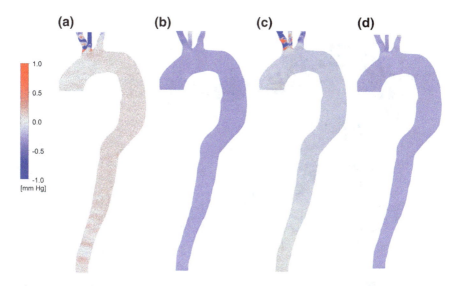

Fig. B.11 Double-stent case pressure differences: medium and fine meshes at **a** peak systole and **b** dicrotic notch, coarse and medium meshes at **c** peak systole and **d** dicrotic notch

Figure B.10 shows the pressure differences between meshes for the single-stent case at peak systole and the dicrotic notch. The largest difference of 1 mmHg can be observed along the dissected region in the fine-medium case (Fig. B.10a). The pressure differences were reduced along the distal TL and show similar variability to the pre-stent case. At the dicrotic notch, the pressure differences were considerably reduced along the domain (Fig. B.10b).

The highest pressure differences between the coarse and medium mesh at peak systole (Fig. B.10c) are seen along the branches and distal TL (≈ -1 mmHg, but varying considerably). The pressure differences are similar at the dicrotic notch (Fig. B.10d).

The pressure differences for the double-stent case are shown in Fig. B.11. It can be seen that the differences are significantly decreased compared to the cases prior to removal of the false lumen, indicating that the sensitivity of the pressure distribution to the mesh refinement is dominated by the flow restriction in the DA caused by the FL.

The highest pressure difference between the medium and fine meshes can be observed at peak systole along the supraaortic branches (Fig. B.11a) and is almost 0 mmHg along the double-stented region. At the dicrotic notch, smaller differences are present along the branches but overall there is a ≈ 0.25 mmHg pressure difference along the double-stent domain (Fig. B.11b). Similar behaviour can be observed when comparing the coarse and medium meshes.

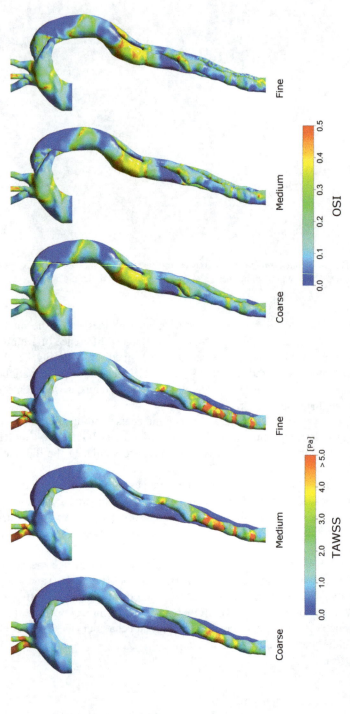

Fig. B.12 TAWSS and OSI for the three meshes for the pre-operative case

Appendix B: Detailed Mesh Sensitivity … 177

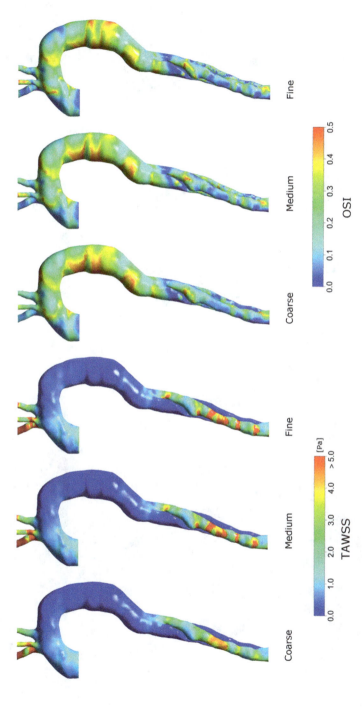

Fig. B.13 TAWSS and OSI for the three meshes for the single-stent case

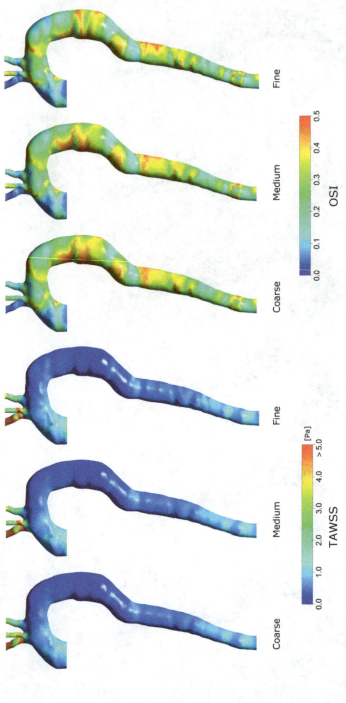

Fig. B.14 TAWSS and OSI for the three meshes for the double-stent case

Appendix B: Detailed Mesh Sensitivity ...

Overall, small variations in local pressure are observed for all the geometries and for all meshes, and these differences tend to be marginally larger when comparing the coarse and medium meshes, rather than the medium and fine meshes. It is clear that the main regions of sensitivity are where the pathological geometry constricts the flow. However, it should also be noted that, relative to the pulse pressure (≈ 50 mmHg), these differences only amount to around $\pm 2\%$. Given the limited resolution of the patient specific data, it is reasonable to conclude that the simulated pressure for this study is insensitive to the mesh size. Furthermore, the calculated flow waves were affected little by the mesh size.

Wall Shear Stress

Finally, the influence of mesh refinement on the wall shear stress indices considered in Chap. 4 are investigated.

Figure B.12 shows the TAWSS and OSI distributions for all three meshes (coarse, medium and fine) for the pre-operative case. The TAWSS in all three mesh cases show a similar distribution in general, however, close inspection shows that the values increase slightly from coarse to medium to fine mesh, for example close to the re-entry tear. Assuming that the fine mesh represents the optimal results, the medium mesh succeeds in identifying regions of elevated TAWSS, but appears to marginally underestimate the magnitude. For the coarse mesh, however, some regions where elevated WSS is predicted in the finer meshes do not show sufficient difference to the surrounding regions to be identified as 'elevated', for example in the coarctation.

For the OSI, similar conclusions can be drawn, in that the magnitude of the OSI was generally increased at a given location as the number of mesh elements increased. Furthermore, more complex distributions can be observed for the fine mesh (consider e.g. the proximal FL). Coarse meshes inevitably report a smoother distribution.

The TAWSS and OSI for all three meshes (coarse, medium and fine) for the single-stent case are shown in Fig. B.13. Application of the stent reduced the TAWSS to very small values in the stented region, and this was the case for all three mesh sizes. Analysing the distal TL, however, the coarse mesh did not yield definitive regions of elevation, while the medium and fine meshes reported more similar results. In the stented region, the OSI was increased by the application of the stent. All three meshes yield similar distributions, although in this case marginally lower values are observed in the fine mesh compared to the coarse mesh.

The TAWSS and OSI for the double-stent case are shown in Fig. B.14. The TAWSS in the double-stent case was almost the same in all three meshes with the exception of the LCC branch, in that the TAWSS becomes higher in the base of LCC branch as the mesh becomes finer. The OSI distribution was almost the same in all three cases between the coarse, medium and fine mesh. See the main text for summary and analysis of these results.

CPSIA information can be obtained
at www.ICGtesting.com
Printed in the USA
LVHW022150060619
620412LV00003B/17/P